SHSAT: POWER PRACTICE

Related Titles

Catholic High School Entrance Exam

SHSAT: POWER PRACTICE

NEW YORK

Library of Congress Cataloging-in-Publication Data:
SHSAT, power practice.—1st ed.

      p.   cm.

  ISBN-13: 978-1-57685-776-2 (pbk. : alk. paper)

  ISBN-10: 1-57685-776-X (pbk. : alk. paper)

  1. Science—Examinations, questions, etc.   2. High schools—New York (State)—New York—Entrance examinations—Study guides.   3. Specialized Science High Schools Admissions Test—Study guides.

  Q182.S57   2011

  373.126'2097471—dc22

                                     2010049437

Printed in the United States of America

9 8 7 6 5 4 3

ISBN 978-1-57685-776-2

For more information or to place an order, contact LearningExpress at:
  2 Rector Street
  26th Floor
  New York, NY 10006

Or visit us at:
  www.learnatest.com

CONTENTS

ABOUT *SHSAT: POWER PRACTICE*

earningExpress understands the importance of achieving top scores on your Specialized High Schools Admissions Test (SHSAT), and we strive to publish the most authentic and comprehensive SHSAT test preparation materials available today. Practice does indeed make perfect, and that's why we've created this book—composed of full-length SHSAT practice exams complete with detailed answer explanations, it offers students all the extra practice they need to get great scores. Whether used on its own or as a powerful companion to other best-selling SHSAT preparation titles, *SHSAT: Power Practice* is the key to a top score and a brighter future.

1 ▶ THE SCHOOLS, THE TEST, AND *SHSAT POWER PRACTICE*

Welcome to *SHSAT Power Practice!* This book is your key to getting an excellent score on New York City's Specialized High School Admissions Test (SHSAT)

If you are planning to take the SHSAT, you already have an idea of your educational goals. You feel that your interests and talents will be best fulfilled in one of New York City's specialized high schools for academically and artistically gifted students. So, you've got your eye on one or more of the following schools:

- **Bronx High School of Science** specializes in science and math, with strong emphasis on humanities and the social sciences.
- **Brooklyn Latin School** offers an in-depth classical education, with a focus on Socratic, student-led discussions.
- **Brooklyn Technical High School** specializes in engineering, the sciences, and computer science, offering hands-on work in laboratories, computer centers, and shops.

- **High School for Mathematics, Science, and Engineering** at the City College specializes in mathematics, engineering, and science, featuring competitions in math and robotics.
- **High School of American Studies** at Lehman College specializes in American history, preparing students for careers in politics, law, journalism, business, science, mathematics, and the arts.
- **Queens High School for the Sciences** at York College specializes in science, mathematics, and technology, offering special opportunities for students seeking a career in medicine.
- **Staten Island Technical High School** features a wide-ranging curriculum of mathematics, science, computers, engineering, and the humanities, offering hands-on work in excellent technical facilities.
- **Stuyvesant High School** specializes in mathematics and offers many extracurricular programs in music and athletics.

The specialized high school not included on this list is **Fiorello H. LaGuardia High School of Music & Art and Performing Arts**, which specializes in the visual and performing arts. Acceptance in this school is determined by academic record and audition or portfolio, not by the SHSAT.

If you haven't already set your academic sights on one or more of the eight schools listed, start thinking about it now. Before you take the SHSAT, you must indicate your school preferences. While you must choose at least one school as your first choice, you will improve your overall chances of acceptance if you rank more than one school. However, do *not* rank any schools that you do not wish to attend. You can learn more about the specialized high schools at the New York City Department of Education website: schools.nyc.gov/Accountability/resources/testing/SHSAT.htm.

The SHSAT

The first thing you should probably know about the SHSAT is that it is highly competitive. In 2010, about 28,000 students took the test. The SHSAT is the only basis for admission into the eight schools previously listed. Even your academic record doesn't count, so it is crucial that you do well on the SHSAT if you want to pursue the academic possibilities that these schools have to offer. This guide will help you perform to the best of your abilities on this demanding exam.

If you are an eighth or ninth grade student in one of New York City's five boroughs, you are ready to take the SHSAT. Most students who take the test are in the eighth grade. You may take it while in both grades if you wish.

SHSAT Test Information

For important SHSAT test information, we suggest you read the *Specialized High Schools Student Handbook*, available on the NYC Department of Education Website (http://schools.nyc.gov).

What's on the Test

The SHSAT consists of 95 multiple choice questions. It is divided into two parts: a verbal section and a math section. The verbal section has 45 questions of the following types:

- 5 scrambled paragraph questions, for which you will be expected to put sentences into their proper sequential order
- 10 logical reasoning questions, which will test your ability to draw logical conclusions from information presented

- 30 reading comprehension questions, which will test your ability to comprehend reading passages of about 350 to 450 words

The mathematics section has 50 questions in the following areas:

- algebra
- arithmetic
- basic coordinate graphing (eighth grade)
- factoring
- geometry (ninth grade)
- logic
- probability
- statistics
- substitution
- word problems

The Time Factor

You will have a total of 150 minutes (two and a half hours) to complete the SHSAT. You may divide your time as you like between the verbal and mathematics sections of the tests, although it is a good idea to allow no more than 75 minutes for each section. You may use a non-calculator watch of your own to keep track of your time.

You may begin with either the mathematics or verbal section. When you complete one section, you may move right on to the other one. When you run into a particularly difficult question, don't spend too much time on it. Either make the best guess you can or come back to it when you have completed the other questions.

If you complete the whole test early, use your remaining time to review answers in both sections. You are not allowed to leave the testing site until the full 150 minutes is up, so make the most of your time.

Test Scoring

You complete your test sheet by filling in bubbles with a #2 pencil. There is only one correct bubble for each test item. If you fill in two bubbles for an item, your answer will be marked as incorrect. Your score is based solely on correct answers. Since there is no penalty for incorrect answers, go ahead and guess whenever you can't figure out the answer to an item. You might want to leave your guesses for last, provided you have time left over when you have completed all the questions you can reliably answer.

Your raw score is based simply on your number of correct answers on the SHSAT. It doesn't mean much for your school acceptance, because individual test forms vary in difficulty. Your raw score will be recalculated as a scale score, which is adjusted according to the difficulty of test forms. It combines your scores in the mathematics and verbal sections. Your scale score determines the school or schools to which you are accepted.

Students are ranked according to their scale scores, from highest to lowest. These rankings are then weighed according to individual school preferences and the number of seats available in each school. When all the seats are filled, a cutoff score is set. If your score is lower than the cutoff score, you will not be accepted into any of the specialized high schools.

Student Notification

You must submit a New York City High School Admissions Application in order to receive notification of your SHSAT results and acceptance. This application becomes available through your counselor in October. The deadline for submission is in December. You will be notified of your score and school offers in February. If you are accepted by more than one school, you have until the end of February to decide which school you wish to attend.

Summer Discovery Program

After you've received your SHSAT results, you may feel that your score is below the cutoff point due to circumstances beyond your control. If so, think about applying to participate in the Summer Discovery Program offered by some of the specialized high schools every year. Performing well in this program may allow you to gain acceptance in a specialized high school despite your low score.

To qualify for the program, you must have scored within a close range of the cutoff score. You must also be able to demonstrate one or more of the following:

- You come from a low-income family and/or attend a school with predominantly low-income students (a Title 1 school).
- You live in a foster home or are a ward of the state.
- You have lived in the United States for no more than four years and/or live in a home where the customary language is not English.

Finally, your school must recommend you as a student with high potential for a specialized high school.

Ask your school counselor how to apply for the Summer Discovery Program.

Getting Ready

All told, there is no better preparation for the SHSAT than doing well in school and keeping your grades up. After all, the SHSAT is an evaluation of what you have learned during your academic career. Still, it is important to be able to mentally access all your years of learning easily and readily. A useful tactic is to list courses you have taken and jot down summaries of what you learned in them. For example, what books did you read in a particular English class? Which books made an impression on you, and why? Review social studies, science, and other courses in a similar manner.

To prepare for the mathematics section of the test, review as much as you can of what you have learned in your math courses over the years. If you have old homework, notes, and tests handy, go over them closely, problem by problem. What have your math strengths been? What have your weaknesses been? Be ready to stretch beyond what you have learned in your classes.

How to Use
SHSAT Power Practice

Chapter 2 of this guide gives you many further useful tips on how to take the SHSAT. Chapters 3 through 8 offer sample tests based on the SHSAT. Keep in mind that the items on the sample tests will *not* be on the actual test. Even so, these tests will help reveal academic strengths and weaknesses that will affect your SHSAT performance. They also give you an opportunity to get psyched up. When working on these practice tests, create an environment and circumstances that closely match those of the SHSAT. Find out how you perform under pressure. Take special care to time your practice sessions and learn how to pace yourself. Only you can decide how much time to take on individual items or on each section.

At the end of this chapter, you will find a sample SHSAT answer sheet. Getting familiar with it is a good way to start preparing. Then make the best use you can of *SHSAT Power Practice*—and good luck on your SHSAT!

2 ▶ THE LEARNINGEXPRESS TEST PREPARATION SYSTEM

Taking any test can be tough, but don't let the Specialized High School Admission Test (SHSAT) scare you. Reviewing the test format and material and learning about your test-taking strengths and weaknesses are keys to achieving a high score on the SHSAT. Make sure you allow enough time to prepare. The LearningExpress Test Preparation System, developed exclusively for LearningExpress by leading test experts, gives you the discipline and attitude you need to accomplish your goals.

First, the bad news: Getting ready for any test takes work. And if you want to go to any of eight New York specialized high schools, you *must* do well on the SHSAT. This book focuses on the verbal and mathematical skills that you will be tested on. By honing these skills, you will take your first step toward acceptance in a specialized high school. However, there are all sorts of pitfalls that can prevent you from doing your best on exams. Here are some obstacles that can stand in the way of your success

- being unfamiliar with the format of the exam
- being paralyzed by test anxiety

- leaving your preparation until the last minute
- not preparing at all
- not knowing vital test-taking skills like:
  - how to pace yourself through the exam
  - how to use the process of elimination
  - when to guess
- not being in tip-top mental and physical shape
- forgetting to eat breakfast and having to take the test on an empty stomach
- forgetting a sweater or jacket and shivering through the exam

What's the common denominator in all these test-taking pitfalls? One word: *control*. Who's in control, you or the exam?

Now the good news: The LearningExpress Test Preparation System puts *you* in control. In just nine easy-to-follow steps, you will learn everything you need to know to make sure you are in charge of your preparation and performance on the exam. *Other* test takers may let the test get the better of them; *other* test takers may be unprepared or out of shape, but not *you*. You will have taken all the steps necessary to score well above the cutoff score.

Here's how the LearningExpress Test Preparation System works: Nine easy steps lead you through everything you need to know and do to get ready to master your exam. While each of the steps listed here gives you tips and activities to help you prepare for any exam, they are specifically geared to help you achieve your goals on the SHSAT. It's important that you follow the advice and do the activities, or you won't be getting the full benefit of the system. Each step gives you an approximate time estimate.

Step 1. Get Information (30 minutes)
Step 2. Conquer Test Anxiety (20 minutes)
Step 3. Make a Plan (50 minutes)
Step 4. Learn to Manage Your Time (10 minutes)
Step 5. Learn to Use the Process of Elimination (20 minutes)
Step 6. Know When to Guess (20 minutes)
Step 7. Reach Your Peak Performance Zone (10 minutes)
Step 8. Get Your Act Together (10 minutes)
Step 9. Do It! (10 minutes)
Total time for the complete system is 180 minutes (3 hours).

Working through the entire system should take you approximately three hours, though it's perfectly okay if you work faster or slower than the time estimates. If you want to take a whole afternoon or evening, you could work through the entire Learning-Express Test Preparation System in one sitting. Otherwise, you can break it up, and do just one or two steps a day for the next several days. It's up to you—remember, *you're* in control.

Step 1: Get Information

Time to complete: 30 minutes
Activities: Read Chapter 1 and the Student Handbook.

Knowledge is power. The first step in the Learning-Express Test Preparation System is finding out everything you can about the types of questions that will be asked on the exam. If you haven't already done so, stop and read Chapter 1 of this book. Also read the *Specialized High Schools Student Handbook*, available for download at schools.nyc.gov/ Accountability/resources/testing/SHSAT.htm.

After you've read Chapter 1 and the Student Handbook, you'll know a lot about the SHSAT, including the overall length of the test, the number and types of questions in the verbal and math sections, and how the test is scored. The Student Handbook also includes two sample tests for you to take.

Step 2: Conquer Test Anxiety

Time to complete: 20 minutes
Activity: Take the Test Stress Test.

Having complete information about the exam is the first step toward getting control of the exam. Next, you have to overcome one of the biggest obstacles to test success—test anxiety. Test anxiety not only impairs your performance on the exam itself, but it can even keep you from preparing. In Step 2, you'll learn stress management techniques that will help you succeed on your exam. Learn these strategies now, and practice them as you work through the tests in this book. They'll be second nature to you by exam day.

Combating Test Anxiety

The first thing you need to know is that a little test anxiety is a good thing. Everyone gets nervous before a big exam—and if that nervousness motivates you to prepare thoroughly, so much the better. If you weren't nervous, that would mean you didn't care about the next long step in your education—but you obviously do because you are reading this book! It's said that Sir Laurence Olivier, one of the foremost British actors of the twentieth century, was ill before every performance. His stage fright didn't impair his work; in fact, it probably gave him a little extra edge—just the kind of edge you need to do well, whether on a stage or in an exam room.

On page 11 is the Test Stress Quiz. Stop here and answer the questions on that page to find out whether your level of test anxiety is something you should worry about.

Stress Management before the Test

If you feel your level of anxiety getting the best of you in the weeks before the test, here is what you need to do to bring the level down.

- **Get prepared.** There's nothing like knowing what to expect. Being prepared will put you in control of test anxiety. That's why you're reading this book. Use it faithfully, and remind yourself that you're better prepared than most of the people taking the test.
- **Practice self-confidence.** A positive attitude is a great way to combat test anxiety. This is no time to be humble or shy. Stand in front of the mirror and say to your reflection, "I'm prepared. I'm full of self-confidence. I'm going to ace this test. I know I can do it." Say it into a tape recorder and play it back once a day. If you hear it often enough, you'll believe it.
- **Fight negative messages.** Every time someone starts telling you how hard the exam is, start repeating your self-confidence messages to them. If the someone with the negative messages is you, telling yourself you don't do well on exams and you just can't do this, don't listen. Turn on your tape recorder and listen to your self-confidence messages.
- **Visualize.** Imagine yourself reporting for your first day at the school you really want to attend. Visualizing success can help make it happen—and it reminds you why you're preparing for the exam so diligently.
- **Exercise.** Physical activity helps calm down your body and focus your mind. Besides, being in good physical shape can actually help you do well on the exam. Go for a run, lift weights, go swimming—and do it regularly.

Stress Management on Test Day

There are several ways you can bring down your level of test anxiety on test day. To find a comfort level, experiment with the following exercises in the weeks before the test, and use the ones that work best for you.

- **Breathe deeply.** Take a deep breath while you count to five, then let it out. Repeat several times.
- **Move your body.** Try rolling your head in a circle. Rotate your shoulders. Shake your hands from the wrist. Many people find these movements very relaxing.
- **Visualize again.** Think of the place where you are most relaxed: lying on the beach in the sun, walking through the park, or sipping a cup of hot tea. Now close your eyes and imagine you're actually there. If you practice in advance, you'll find that you need only a few seconds of this exercise to experience a significant increase in your sense of well-being.

When anxiety threatens to overwhelm you right there during the exam, there are still things you can do to manage your stress level.

- **Repeat your self-confidence messages.** You should have them memorized by now. Say them quietly to yourself, and believe them!

- **Visualize one more time.** This time, visualize yourself moving smoothly and quickly through the test, answering every question right and finishing just before time is up. Like most visualization techniques, this one works best if you've practiced it ahead of time.
- **Find an easy question.** Skim over the test until you find an easy question, and then answer it. Filling in even one circle gets you into the test-taking groove.
- **Take a mental break.** Everyone loses concentration once in a while during a long test. It's normal, so you shouldn't worry about it. Instead, accept what has happened. Say to yourself, "Hey, I lost it there for a minute. My brain is taking a break." Put down your pencil, close your eyes, and do some deep breathing for a few seconds. Then you're ready to go back to work.

Try these techniques ahead of time, and see if they work for you.

TEST STRESS QUIZ

You need to worry about test anxiety only if it is extreme enough to impair your performance. The following questionnaire will provide a diagnosis of your level of test anxiety. In the blank before each statement, write the number that most accurately describes your experience.

0 = Never
1 = Once or twice
2 = Sometimes
3 = Often

_____ I have gotten so nervous before an exam that I put down the books and did not study for it.

_____ I have experienced disabling physical symptoms such as vomiting and severe headaches because I was nervous about an exam.

_____ I have simply not shown up for an exam because I was afraid to take it.

_____ I have experienced dizziness and disorientation while taking an exam.

_____ I have had trouble filling in the little circles because my hands were shaking too hard.

_____ I have failed an exam because I was too nervous to complete it.

_____ **Total: Add up the numbers in the blanks above.**

Your Test Stress Score

Here are the steps you should take, depending on your score. If you scored:

- **Below 3:** Your level of test anxiety is nothing to worry about; it is probably just enough to give you that little extra edge.
- **Between 3 and 6:** Your test anxiety may be enough to impair your performance, and you should practice the stress management techniques in this section to try to bring your test anxiety down to manageable levels.
- **Above 6:** Your level of test anxiety is a serious concern. In addition to practicing the stress management techniques listed in this section, you may want to seek additional, personal help. Call your local high school or community college and ask for the academic counselor. Tell the counselor that you have a level of test anxiety that sometimes keeps you from being able to take the exam. The counselor may be willing to help you or may suggest someone else you should talk to.

Step 3: Make a Plan

Time to complete: 50 minutes
Activity: Construct a study plan.

Maybe the most important thing you can do to get control of yourself and your exam is to make a study plan. Too many people fail to prepare simply because they fail to plan. Spending hours on the day before the exam poring over sample test questions not only raises your level of test anxiety, it is also no substitute for careful preparation and practice.

Don't fall into the cram trap. Take control of your preparation time by mapping out a study schedule. If you're the kind of person who needs deadlines and assignments to motivate you for a project, here they are. If you're the kind of person who doesn't like to follow other people's plans, you can use the suggested schedules here to construct your own.

Even more important than making a plan is making a commitment. You can't review everything you need to know for the SHSAT in one night. You have to set aside some time every day for study and practice. Try for at least 20 minutes a day. Twenty minutes daily will do you much more good than two hours on Saturday.

Don't put off studying until the day or week before the exam. Start now. Even ten minutes a day, with half an hour or more on weekends, can make a big difference in your score and in your chances of making the grade you want.

Schedule A: The 30-Day Plan

If you have at least one month before you take your test, you have plenty of time to prepare—as long as you don't procrastinate! If you have less than a month, turn to Schedule B.

> **Day 1:** Carefully review Chapter 1 of this book and the *Specialized High Schools Student Handbook* that you read or downloaded in Step 1. Make sure that the contents of the SHSAT are fresh in your mind.

> **Day 2:** Take the first sample test (Chapter 3) and score yourself. Be sure to take the test during one 150-minute session. If you cannot finish the test on time, fill in the remaining answers with guesses, just as you would on the actual SHSAT. Your goal is to make your practice experiences as much like the real test as possible.

> **Day 3:** Review your test performance. Which items did you have to guess at? Which items did you get wrong even when you thought you knew the answers? Did you finish the test on time, or did you have to make guesses on later questions? If you finished the test early, how well did you use your remaining time? Go over all your wrong answers and make sure you understand them.

> **Days 4 and 5:** Now that you have a good idea of what you're up against with the SHSAT, start studying! Hit the books on days 4 and 5.

Reading a lot is essential preparation for the verbal section of the test—books, stories, articles, essays, anything that increases your general knowledge and thinking skills. Read actively, not passively. Ask yourself questions about the author's purpose, what information the author conveys, and what kind of vocabulary the author uses and why. Finally, ask yourself about the effectiveness of the writing. Does it fulfill the author's purpose? If so, how? If not, why not?

Download pertinent math curricula from NYSED at www.emsc.nysed.gov/ciai/pub/pubmath.html. Solving five to ten problems daily is excellent exercise. Working on familiar problems will reinforce your math skills, and working on unfamiliar problems will increase your skills, allowing you to perform on the SHSAT beyond what you have actually learned in school.

Days 6 and 7: Repeat your regimen of days 2 and 3, taking the second test (Chapter 4) on day 6 and reviewing your performance on day 7.

Days 8 and 9: Repeat your regimen of days 4 and 5, reading and doing math problems.

Days 10 through 25: Repeat the four-day routine used on days 2 through 5 and days 6 through 9 over a 16-day period; complete Test 6 (Chapter 8) on days 22 and 23; use days 24 and 25 to study.

Day 26: Take the first sample test in the *Specialized High Schools Student Handbook.*

Day 27: Review your test exactly as you did with the sample tests in *SHSAT Power Prep.*

Days 28 and 29: Repeat your activities on days 26 and 27 with the second sample test in the *Specialized High Schools Student Handbook.*

Day before the exam: Relax. Do something unrelated to the exam and go to bed at a reasonable hour.

Schedule B: The 14-Day Plan

If only you have two weeks, you have your work cut out for you. Use this 14-day schedule to help you make the most of your time.

Day 1: Review Chapter 1 and the Student Handbook.

Day 2: Take and review the first sample test (Chapter 3).

Day 3: Take and review the second sample test (Chapter 4).

Day 4: Read books, stories, articles, essays; practice math problems.

Days 5 and 6: Take and review the third and fourth sample tests (Chapters 5 and 6).

Days 7: Read books, stories, articles, essays; practice math problems.

Day 8 and 9: Take and review the fifth and sixth sample tests (Chapters 7 and 8).

Day 10: Read books, stories, articles, essays; practice math problems.

Day 11: Take and review the first sample test in the *Specialized High Schools Student Handbook.*

Day 12: Take and review the second sample test in the Student Handbook.

Day 13: Read books, stories, articles, essays; practice math problems.

Day before the exam: Relax. Do something unrelated to the exam and go to bed at a reasonable hour.

Step 4: Learn to Manage Your Time

Time to complete: 10 minutes to read, many hours of practice!
Activities: Use these strategies as you take the sample tests in this book.

Steps 4, 5, and 6 of the LearningExpress Test Preparation System put you in charge of your exam by showing you test-taking strategies that work. Practice these strategies as you take the sample tests in this book, and then you'll be ready to use them on test day.

First, take control of your time on the exam. The SHSAT has a time limit of 150 minutes, which may give you more than enough time to complete all the questions—or not enough time. It's a terrible feeling to hear the examiner say, "Five minutes left," when you're only three quarters of the way through the test. Here are some tips to keep that from happening to you.

■ **Follow directions.** If the directions are given orally, listen closely. If they're written in the exam booklet, read them carefully. If there is anything you don't understand, ask questions *before* the exam begins. You are allowed to write in your exam booklet, so write down the beginning time and ending time of the exam.

- **Pace yourself.** Glance at your watch every few minutes and compare the time to how far you've gotten in the test. When 50 minutes have elapsed, you should be about a third of the way through the test (or two-thirds of the way through either the math or verbal section), and so on. If you're falling behind, pick up the pace a bit.

- **Keep moving.** Don't waste time on one question. If you don't know the answer, skip the question and move on. Circle the number of the question in your test booklet in case you have time to come back to it later.

- **Keep track of your place on the answer sheet.** If you skip a question, make sure you skip it on the answer sheet, too. Check yourself every 5 or 10 questions to make sure the question number and the answer sheet number are still the same.

- **Don't rush.** Although you should keep moving, rushing won't help. Try to keep calm and work methodically and quickly.

Step 5: Learn to Use the Process of Elimination

Time to complete: 20 minutes
Activity: Complete the worksheet on using the process of elimination.

After time management, your most important tool for taking control of your exam is using the process of elimination wisely. It's standard test-taking wisdom that you should always read all the answer choices before choosing your answer. This helps you find the right answer by eliminating wrong answer choices. And, sure enough, that standard wisdom applies to the SHSAT, too.

Choosing the Right Answer by Process of Elimination

As you read a question, you may find it helpful to underline important information or make some notes about what you're reading. When you get to the heart of the question, circle it and make sure you understand what it is asking. If you're not sure of what's being asked, you'll never know whether you've chosen the right answer. What you do next depends on the type of question you're answering.

- Take a quick look at the answer choices for some clues. Sometimes this helps to put the question in a new perspective and makes it easier to answer. Then make a plan of attack to solve the problem.

- Otherwise, follow this simple process-of-elimination plan to manage your testing time as efficiently as possible: Read each answer choice and make a quick decision about what to do with it, marking your test book accordingly:
  - ✔ The answer seems reasonable; keep it. Put a ✔ next to the answer.
  - ✔ The answer is awful. Get rid of it. Put an **X** next to the answer.
  - ✔ You can't make up your mind about the answer, or you don't understand it. Keep it for now. Put a **?** next to it.

Whatever you do, don't waste time with any one answer choice. If you can't figure out what an answer choice means, don't worry about it. If it's the right answer, you'll probably be able to eliminate all the others; if it's the wrong answer, another answer choice will probably strike you more obviously as the right answer.

- If you haven't eliminated any answers at all, skip the question temporarily, but don't forget to mark the question so you can come back to it later if you have time. Because the SHSAT has no penalty for wrong answers, if you're certain that you could never answer this question in a million years, pick an answer and move on.

- If you've eliminated all but one answer, just reread the circled part of the question to be sure you're answering exactly what's asked. Mark your answer sheet and move on to the next question.
- If you've eliminated some—but not all—of the answer choices, compare the remaining answers, looking for similarities and differences, reasoning your way through these choices. Try to eliminate those choices that don't seem as strong to you. But *don't* eliminate an answer just because you don't understand it. You may even be able to use relevant information from other parts of the test. If you're down to only two or three answer choices, you've improved your odds of getting the question right. Make an educated guess and move on. However, if you think you can do better with more time, mark the question as one to return to later.

Guess on Every Question

You will *not* be penalized for getting a wrong answer on the SHSAT. This is very good news. That means you should absolutely answer every single question on the test. If you're hopelessly lost on a question and can't even cross off one answer choice, make sure that you don't leave it blank. Even if you only have 30 seconds left and 10 questions still to answer, you should just guess on all those last questions.

Of course, if you can eliminate even one of the choices, you improve your odds of guessing. If you

can identify *two* of the choices as definitely wrong, you have a one in two chance of answering the question correctly. Either way, be sure to answer every question.

If You Finish Early

Use any time you have left to do the following:

- Go back to questions you marked to return to later and try them again.
- Check your work on all the other questions. If you have a good reason for thinking a response is wrong, change it.
- Review your answer sheet. Make sure you've put the answers in the right places and you've marked only one answer for each question. Remember, if you mark more than one answer on an SHSAT item, that item will be marked wrong.
- If you've erased an answer, make sure you've done a good job of it.
- Check for stray marks on your answer sheet that could distort your score.

Whatever you do, don't waste time when you've finished a test section. Make every second count by checking your work over and over again until time is called. Try using your powers of elimination on the questions in the worksheet that follows called "Using the Process of Elimination." The answer explanations that follow show possible methods for arriving at the right answer.

Process of elimination is your tool for the next step, which is knowing when to guess.

Use the process of elimination to answer the following questions.

1. Ilsa is as old as Meghan will be in five years. The difference between Ed's age and Meghan's age is twice the difference between Ilsa's age and Meghan's age. Ed is 29. How old is Ilsa?
   a. 4
   b. 10
   c. 19
   d. 24

2. "All drivers of commercial vehicles must carry a valid commercial driver's license whenever operating a commercial vehicle." According to this sentence, which of the following people need NOT carry a commercial driver's license?
   a. a truck driver idling his engine while waiting to be directed to a loading dock
   b. a bus operator backing her bus out of the way of another bus in the bus lot
   c. a taxi driver driving his personal car to the grocery store
   d. a limousine driver taking the limousine to her home after dropping off her last passenger of the evening

3. Smoking tobacco has been linked to
   a. increased risk of stroke and heart attack.
   b. all forms of respiratory disease.
   c. increasing mortality rates over the past ten years.
   d. juvenile delinquency.

4. Which of the following words is spelled correctly?
   a. incorrigible
   b. outragous
   c. domestickated
   d. understandible

Answers

Here are the answers, as well as some suggestions as to how you might have used the process of elimination to find them.

1. **d.** You should have eliminated choice **a** right off the bat. Ilsa cannot be four years old if Meghan is going to be Ilsa's age in five years. The best way to eliminate other answer choices is to try plugging them in to the information given in the problem. For instance, for choice **b**, if Ilsa is 10, then Meghan must be 5. The difference between their ages is 5. The difference between Ed's age, 29, and Meghan's age, 5, is 24. Is 24 two times 5? No. Then choice **b** is wrong. You could eliminate choice **c** in the same way and be left with choice **d**.

2. **c.** Note the word *NOT* in the question, and go through the answers one by one. Is the truck driver in choice **a** operating a commercial vehicle? Yes, idling counts as operating, so he needs to have a commercial driver's license. Likewise, the bus operator in choice **b** is operating a commercial vehicle; the question doesn't say the operator has to be on the street. The limo driver in choice **d** is operating a commercial vehicle, even if it doesn't have a passenger in it. However, the driver in choice **c** is not operating a commercial vehicle, but his own private car.

3. a. You could eliminate choice **b** simply because of the presence of the word *all*. Such absolutes hardly ever appear in correct answer choices. Choice **c** looks attractive until you think a little about what you know—aren't fewer people smoking these days, rather than more? So how could smoking be responsible for a higher mortality rate? (If you didn't know that mortality rate means the rate at which people die, you might keep this choice as a possibility, but you would still be able to eliminate two answers and have only two to choose from.) And choice **d** is plain silly, so you could eliminate that one, too. You are left with the correct choice, **a**.

4. a. How you used the process of elimination here depends on which words you recognized as being spelled incorrectly. If you knew that the correct spellings were outrageous, domesticated, and understandable, then you were home free. If you knew the correct spelling of one or two of these words, you could improve your chances of guessing correctly by using elimination. Surely you knew that at least one of those words was wrong!

Step 6: Know When to Guess

Time to complete: 20 minutes
Activity: Complete worksheet on Your Guessing Ability.

Armed with the process of elimination, you're ready to take control of one of the big questions in test taking: Should I guess? In the SHSAT, the number of questions you answer correctly yields your raw score, which is used to determined your scaled score. So you have nothing to lose and everything to gain by guessing.

The more complicated answer to the question, "Should I guess?" depends on you, your personality, and your guessing intuition. There are two things you need to know about yourself before you go into the exam:

1. Are you a risk taker?
2. Are you a good guesser?

You'll have to decide about your risk-taking quotient on your own. To find out whether you're a good gueser, complete the worksheet called "Your Guessing Ability" that begins on page 18. Frankly, even if you're a play-it-safe person with terrible intuition, you're still safe in guessing every time, because the exam has no guessing penalty. It would be best for you to overcome your anxieties and go ahead and mark an answer. But you may want to have a sense of how good your intuition is before you go into the exam.

YOUR GUESSING ABILITY

The following are ten really hard questions. You're not supposed to know the answers. Rather, this is an assessment of your ability to guess when you don't have a clue. Read each question carefully, just as if you did expect to answer it. If you have any knowledge at all of the subject of the question, use that knowledge to help you eliminate wrong answer choices. Use this answer grid to fill in your answers to the questions. Bear in mind that while these questions do not resemble those you will find on the SHSAT, they are a good indication of your ability to make an educated guess.

1. (a) (b) (c) (d)   5. (a) (b) (c) (d)   9. (a) (b) (c) (d)
2. (a) (b) (c) (d)   6. (a) (b) (c) (d)   10. (a) (b) (c) (d)
3. (a) (b) (c) (d)   7. (a) (b) (c) (d)
4. (a) (b) (c) (d)   8. (a) (b) (c) (d)

1. September 7 is Independence Day in
   a. India.
   b. Costa Rica.
   c. Brazil.
   d. Australia.

2. Which of the following is the formula for determining the momentum of an object?
   a. $p = mv$
   b. $F = ma$
   c. $P = IV$
   d. $E = mc^2$

3. Because of the expansion of the universe, the stars and other celestial bodies are all moving away from each other. This phenomenon is known as
   a. Newton's first law.
   b. the big bang.
   c. gravitational collapse.
   d. Hubble flow.

4. American author Gertrude Stein was born in
   a. 1713.
   b. 1830.
   c. 1874.
   d. 1901.

5. Which of the following is NOT one of the Five Classics attributed to Confucius?
   a. the *I Ching*
   b. the *Book of Holiness*
   c. the *Spring and Autumn Annals*
   d. the *Book of History*

6. The religious and philosophical doctrine that holds that the universe is constantly in a struggle between good and evil is known as
   a. Pelagianism.
   b. Manichaeanism.
   c. neo-Hegelianism.
   d. Epicureanism.

7. The third Chief Justice of the U.S. Supreme Court was
   a. John Blair.
   b. William Cushing.
   c. James Wilson.
   d. John Jay.

8. Which of the following is the poisonous portion of a daffodil?
   a. the bulb
   b. the leaves
   c. the stem
   d. the flowers

9. The winner of the Masters golf tournament in 1953 was
   a. Sam Snead.
   b. Cary Middlecoff.
   c. Arnold Palmer.
   d. Ben Hogan.

10. The state with the highest per-capita personal income in 1980 was
    a. Alaska.
    b. Connecticut.
    c. New York.
    d. Texas.

Answers

Check your answers against the correct answers below.

1. c
2. a
3. d
4. c
5. b
6. b
7. b
8. a
9. d
10. a

How Did You Do?

You may have simply gotten lucky and actually known the answer to one or two questions. In addition, your guessing was more successful if you were able to use the process of elimination on any of the questions. Maybe you did not know who the third Chief Justice was (question 7), but you knew that John Jay was the first. In that case, you would have eliminated answer **d**

and, therefore, improved your odds of guessing right from one in four to one in three.

According to probability, you should get two and a half answers correct, so getting either two or three right would be average. If you got four or more right, you may be a really terrific guesser. If you got one or none right, you may have decided not to guess. Remember not to leave any question blank, no matter how hard it may seem!

You should continue to keep track of your guessing ability as you work through the sample tests in this book. Circle the numbers of questions you guess at; or, if you don't have time during the practice tests, go back afterward and try to remember which answers were guesses. Remember, on a test with four answer choices, your chances of getting a right answer is one in four. So keep a separate guessing score for each exam. How many of your answers were guesses? How many did you get right? If the number you got right is at least one-fourth of the number of questions you guessed at, you are at least an average guesser—and you should always go ahead and guess on the real exam.

Step 7: Reach Your Peak Performance Zone

Time to complete: 10 minutes to read; weeks to complete!

Activity: Complete the Physical Preparation Checklist

To get ready for a challenge like a big exam, you have to take control of your physical as well as your mental state. Exercise, proper diet, and rest will ensure that your body works with, rather than against, your mind on test day, as well as during your preparation.

Exercise

If you don't already have a regular exercise program going, the time during which you're preparing for an exam is actually an excellent time to start one. If you're already keeping fit—or trying to get that way—don't let the pressure of preparing for an exam fool you into quitting now. Exercise helps reduce stress by pumping wonderful good-feeling hormones called endorphins into your system. It also increases the oxygen supply throughout your body and your brain, so you'll be at peak performance on test day.

A half hour of vigorous activity—enough to break a sweat—every day should be your aim. If you're really pressed for time, every other day is okay. Choose an activity you like and get out there and do it. Jogging with a friend always makes the time go faster, as does listening to music.

But don't overdo it. You don't want to exhaust yourself. Moderation is the key.

Diet

First of all, cut out the junk. Go easy on caffeine, and promise yourself a special treat the night after the exam, if need be.

Your body needs a balanced diet for peak performance. Eat plenty of fruits and vegetables, along with protein complex and carbohydrates. Foods that are high in lecithin (an amino acid), such as fish and beans, are especially good "brain foods."

Rest

You probably know how much sleep you need every night to be at your best, even if you don't always get it. Make sure you do get that much sleep, though, for at least a week before the exam. Moderation is important here, too. Extra sleep will just make you groggy.

If you're not a morning person and your exam will be given in the morning, you should reset your internal clock so that your body doesn't think you're taking an exam at 3 A.M. You have to start this process well before the exam. The way it works is to get up half an hour earlier each morning, and then go to bed half an hour earlier that night. Don't try it the other way around; you'll just toss and turn if you go to bed early without getting up early. The next morning, get up another half an hour earlier, and so on. How long you will have to do this depends on how late you're used to getting up. Use the "Physical Preparation Checklist" on page 21 to make sure you're in tip-top form.

Physical Preparation Checklist

For the week before the test, write down what physical exercise you engaged in and for how long, and what you ate for each meal. Remember, you're trying for at least half an hour of exercise every other day (preferably every day) and a balanced diet that's light on junk food.

For the week before the test, write down what physical exercise you engaged in and for how long, and what you ate for each meal. Remember, you are trying for at least half an hour of exercise every other day (preferably every day) and a balanced diet that is light on junk food.

Exam minus 7 days

Exercise: _____ for _____ minutes

Breakfast: _____

Lunch: _____

Dinner: _____

Snacks: _____

Exam minus 6 days

Exercise: _____ for _____ minutes

Breakfast: _____

Lunch: _____

Dinner: _____

Snacks: _____

Exam minus 5 days

Exercise: _____ for _____ minutes

Breakfast: _____

Lunch: _____

Dinner: _____

Snacks: _____

Exam minus 4 days

Exercise: _____ for _____ minutes

Breakfast: _____

Lunch: _____

Dinner: _____

Snacks: _____

Exam minus 3 days

Exercise: _____ for _____ minutes

Breakfast: _____

Lunch: _____

Dinner: _____

Snacks: _____

Exam minus 2 days

Exercise: _____ for _____ minutes

Breakfast: _____

Lunch: _____

Dinner: _____

Snacks: _____

Exam minus 1 day

Exercise: _____ for _____ minutes

Breakfast: _____

Lunch: _____

Dinner: _____

Snacks: _____

Step 8: Get Your Act Together

Time to complete: 10 minutes to read; time to complete will vary
Activity: Complete the Final Preparations Worksheet.

Once you feel in control of your mind and body, you're in charge of test anxiety, test preparation, and test-taking strategies. Now it's time to make charts and gather the materials you need to take to the exam.

Gather Your Materials

The night before the exam, lay out the clothes you will wear and the materials you have to bring with you to the exam. Plan on dressing in layers because you won't have any control over the temperature of the exam room. Have a sweater or jacket you can take off if it's warm. Use the checklist on the worksheet entitled "Final Preparations" on page 23 to help you pull together what you'll need.

Follow Your Routine

If you usually have coffee and toast every morning, then you should have coffee and toast before the test. If you don't usually eat breakfast, don't start changing your habits on exam morning. Do whatever you normally do so that your body will be used to it. If you're not used to it, a cup of coffee can really disrupt your stomach. Doughnuts or other sweet foods can give you a stomachache, too. When deciding what to have for breakfast, remember that a sugar high will leave you with a sugar low in the middle of the exam. A mix of protein and carbohydrates is best: Cereal with milk or eggs with toast will do your body a world of good.

Getting to the Exam Site

Location of exam site: _____

Date: ___ _____

Departure time: _____

Do I know how to get to the exam site?   Yes _____   No _____   (If no, make a trial run.)

Time it will take to get to exam site _____

Things to Lay Out the Night Before

Clothes I will wear   _____

Sweater/jacket   _____

Watch   _____

Photo ID   _____

Four #2 pencils   _____

Other Things to Bring/Remember

_____   _____

_____   _____

_____   _____

_____   _____

_____   _____

Step 9: Do It!

**Time to complete: 10 minutes, plus test-taking time
Activity: Ace Your Test!**

Fast-forward to exam day. You're ready. You made a study plan and followed through. You practiced your test-taking strategies while working through this book. You're in control of your physical, mental, and emotional state. You know when and where to show up and what to bring with you. In other words, you're better prepared than most of the other people taking the test. You're psyched!

Just one more thing. When you're finished with the exam, you will have earned a reward. Plan a night out. Call your friends and plan a get-together, or have a nice dinner with your family to celebrate—whatever your heart desires. Give yourself something to look forward to.

And then do it. Go into the exam, full of confidence, armed with test-taking strategies you've practiced until they're second nature. You're in control of yourself, your environment, and your performance on exam day. You're ready to succeed. So do it. Go in there and ace the SHSAT! And then look forward to your years in a specialized high school.

C H A P T E R

▶ PRACTICE TEST 1

The *SHSAT Power Practice* tests will help you prepare for the high stakes exams given to students applying for New York City's specialized high schools. Each practice test consists of sample questions like those you will find on the official SHSAT.

The 45 question verbal section and 50 question math section were developed by education experts. These tests will show you how much you know and what kinds of problems you still need to study. Mastering these practice tests will allow you to reach your highest potential on the real SHSAT.

PART I VERBAL

Scrambled Paragraphs

Paragraph 1

(q) (r) (s) (t) (u)
(q) (r) (s) (t) (u)
(q) (r) (s) (t) (u)
(q) (r) (s) (t) (u)
(q) (r) (s) (t) (u)

Paragraph 2

(q) (r) (s) (t) (u)
(q) (r) (s) (t) (u)
(q) (r) (s) (t) (u)
(q) (r) (s) (t) (u)
(q) (r) (s) (t) (u)

Paragraph 3

(q) (r) (s) (t) (u)
(q) (r) (s) (t) (u)
(q) (r) (s) (t) (u)
(q) (r) (s) (t) (u)
(q) (r) (s) (t) (u)

Paragraph 4

(q) (r) (s) (t) (u)
(q) (r) (s) (t) (u)
(q) (r) (s) (t) (u)
(q) (r) (s) (t) (u)
(q) (r) (s) (t) (u)

Paragraph 5

(q) (r) (s) (t) (u)
(q) (r) (s) (t) (u)
(q) (r) (s) (t) (u)
(q) (r) (s) (t) (u)
(q) (r) (s) (t) (u)

Logical Reasoning

6. (a) (b) (c) (d) (e)
7. (f) (g) (h) (j) (k)
8. (a) (b) (c) (d) (e)
9. (f) (g) (h) (j) (k)
10. (a) (b) (c) (d) (e)
11. (f) (g) (h) (j) (k)
12. (a) (b) (c) (d) (e)
13. (f) (g) (h) (j) (k)
14. (a) (b) (c) (d) (e)
15. (f) (g) (h) (j) (k)

Reading

16. (a) (b) (c) (d) (e)
17. (f) (g) (h) (j) (k)
18. (a) (b) (c) (d) (e)
19. (f) (g) (h) (j) (k)
20. (a) (b) (c) (d) (e)
21. (f) (g) (h) (j) (k)
22. (a) (b) (c) (d) (e)
23. (f) (g) (h) (j) (k)
24. (a) (b) (c) (d) (e)
25. (f) (g) (h) (j) (k)
26. (a) (b) (c) (d) (e)
27. (f) (g) (h) (j) (k)
28. (a) (b) (c) (d) (e)
29. (f) (g) (h) (j) (k)
30. (a) (b) (c) (d) (e)

31. (f) (g) (h) (j) (k)
32. (a) (b) (c) (d) (e)
33. (f) (g) (h) (j) (k)
34. (a) (b) (c) (d) (e)
35. (f) (g) (h) (j) (k)
36. (a) (b) (c) (d) (e)
37. (f) (g) (h) (j) (k)
38. (a) (b) (c) (d) (e)
39. (f) (g) (h) (j) (k)
40. (a) (b) (c) (d) (e)
41. (f) (g) (h) (j) (k)
42. (a) (b) (c) (d) (e)
43. (f) (g) (h) (j) (k)
44. (a) (b) (c) (d) (e)
45. (f) (g) (h) (j) (k)

PART II MATHEMATICS

46. (a) (b) (c) (d) (e)
47. (f) (g) (h) (j) (k)
48. (a) (b) (c) (d) (e)
49. (f) (g) (h) (j) (k)
50. (a) (b) (c) (d) (e)
51. (f) (g) (h) (j) (k)
52. (a) (b) (c) (d) (e)
53. (f) (g) (h) (j) (k)
54. (a) (b) (c) (d) (e)
55. (f) (g) (h) (j) (k)
56. (a) (b) (c) (d) (e)
57. (f) (g) (h) (j) (k)
58. (a) (b) (c) (d) (e)
59. (f) (g) (h) (j) (k)
60. (a) (b) (c) (d) (e)
61. (f) (g) (h) (j) (k)
62. (a) (b) (c) (d) (e)

63. (f) (g) (h) (j) (k)
64. (a) (b) (c) (d) (e)
65. (f) (g) (h) (j) (k)
66. (a) (b) (c) (d) (e)
67. (f) (g) (h) (j) (k)
68. (a) (b) (c) (d) (e)
69. (f) (g) (h) (j) (k)
70. (a) (b) (c) (d) (e)
71. (f) (g) (h) (j) (k)
72. (a) (b) (c) (d) (e)
73. (f) (g) (h) (j) (k)
74. (a) (b) (c) (d) (e)
75. (f) (g) (h) (j) (k)
76. (a) (b) (c) (d) (e)
77. (f) (g) (h) (j) (k)
78. (a) (b) (c) (d) (e)
79. (f) (g) (h) (j) (k)

80. (a) (b) (c) (d) (e)
81. (f) (g) (h) (j) (k)
82. (a) (b) (c) (d) (e)
83. (f) (g) (h) (j) (k)
84. (a) (b) (c) (d) (e)
85. (f) (g) (h) (j) (k)
86. (a) (b) (c) (d) (e)
87. (f) (g) (h) (j) (k)
88. (a) (b) (c) (d) (e)
89. (f) (g) (h) (j) (k)
90. (a) (b) (c) (d) (e)
91. (f) (g) (h) (j) (k)
92. (a) (b) (c) (d) (e)
93. (f) (g) (h) (j) (k)
94. (a) (b) (c) (d) (e)
95. (f) (g) (h) (j) (k)

Part 1—Verbal

The Verbal Test includes 45 questions in these three sections:

- Scrambled Paragraphs, 5 paragraphs (each counts double)
- Logical Reasoning, 10 questions, numbered 6–15
- Reading, 30 questions, numbered 16–45

Scrambled Paragraphs

This section tests your ability to organize a paragraph well. There are five paragraphs, presented in scrambled order. Your job is to put them in the best order to make a clear, coherent paragraph. Each correct answer counts double; these five paragraphs are worth 10 points out of the 50-point verbal test.

The first sentence in each paragraph is given. The remaining five sentences are listed in random order. Read each group of sentences carefully, and then decide on the best arrangement for them. Use the blanks at the left of each sentence to number these sentences from 1 to 5, showing the order they should be in.

Paragraph 1
Cupid struck Apollo with his enchanted arrows, and Apollo fell in love with a nymph called Daphne.

_____ **Q.** The river god transformed Daphne into a tree, and she instantly grew roots, bark, and leaves.

_____ **R.** Apollo chased Daphne through the forest.

_____ **S.** Daphne prayed to her father, the river god, for help.

_____ **T.** Apollo decided that if the nymph could not be his wife, he would honor her by using her boughs for his crown and wreaths.

_____ **U.** Apollo confessed his love to Daphne, but she ignored him and began to run away.

Paragraph 2
Goran Kropp was a famous Swedish mountaineer.

_____ **Q.** Goran Kropp's amazing journey did not end there; when he came down from Everest, he got back on his bike and rode the 7,000 miles home.

_____ **R.** Twenty-three years later, he did something that most people said was impossible.

_____ **S.** He climbed his first mountain when he was just six years old.

_____ **T.** But that was just the start; then Kropp climbed the 29,000-foot mountain without any help, carrying his own supplies on his back.

_____ **U.** He rode his bike from his home in Sweden, 7,000 miles to the highest mountain in the world, Mt. Everest.

Paragraph 3
If your office job involves telephone work, you are in a very important position.

_____ **Q.** After a short, friendly greeting, state your company's name and then your own name.

_____ **R.** Your faceless voice may be the first contact a caller has with your company or organization.

_____ **S.** For this reason, your telephone manners have to be impeccable.

_____ **T.** Always take messages carefully, and always let the caller hang up first.

_____ **U.** Answer the phone promptly, on the first or second ring if possible.

Paragraph 4

Although it may look easy to those who have never practiced, yoga requires great concentration.

_____ **Q.** A standing pose known as *brave warrior* is even more difficult than *staff pose.*

_____ **R.** For example, consider a simple sitting pose such as *staff pose.*

_____ **S.** Even so, yoga poses are worth the effort because they are surprisingly effective in stretching and strengthening muscles.

_____ **T.** It requires you to balance on one leg and hold a pose that strengthens leg, back, and stomach muscles.

_____ **U.** As you sit, the pose requires you to tighten and lengthen stomach, back, and arm muscles as you stretch your legs out in front of you and place your hands by your sides.

Paragraph 5

Plan your holiday party early.

_____ **Q.** Be sure to include the date, time, and location on your invitations.

_____ **R.** Planning early also gives you time to send out invitations and get replies so you will know how many people to expect.

_____ **S.** You may also want to include special instructions on the invitation if you expect your guests to bring something or if you want to have a themed party.

_____ **T.** One reason to get started early is that people make their own plans for popular holidays such as Memorial Day and Labor Day.

_____ **U.** If you don't decide until two weeks before one of these holidays that you want to have a party, you may not have many people show up.

Logical Reasoning

The questions in this section test your ability to reason well, that is, to figure out what the facts you know can or can't possibly mean. Read the statements carefully, then choose the best answer based *only* on the information given. Note carefully the words used in each question. For example, one thing can be larg*er* than another without being the larg*est* in the group. In answering some of these questions, it may be useful to draw a rough diagram or make a list that gives real-world values to the information.

6. Four eighth graders were sharing a pizza. They decided that the oldest among them would get the extra piece.

Randy is two months older than Greg.
Greg is three months younger than Ned.
Kent is one month older than Greg.

Who should get the extra piece of pizza?

a. Randy
b. Greg
c. Ned
d. Kent
e. It cannot be determined from the information given.

7. All dogs like to run. Some dogs like to swim. Some dogs look like their masters.

Read the following statements. Based on the preceding information, which statement or statements *must* be true?

　I. Dogs who like to run look like their masters.
　II. Dogs who like to swim like to run.
　III. Dogs who look like their masters like to run.

f. II only
g. III only
h. I and II only
j. II and III only
k. I, II, and III

8. Over summer vacation, Mark and Consuela took their car on a road trip to Mexico City. Mark spent 48 hours behind the wheel. Consuela spent 36 hours driving.

Which of the following can we conclude is true?

  a. Mark and Consuela are excellent drivers.
  b. Consuela spent less time driving the car than Mark.
  c. Mark drove below the speed limit.
  d. Mark is a more experienced driver than Consuela.
  e. Consuela doesn't like to drive.

9. Megan never drinks her coffee black. Sometimes she has sugar in it, but she always has cream. She had three cups of coffee today.

Based on the preceding statements, which of the following must be true?

  f. Megan had sugar today.
  g. Megan does not like black coffee.
  h. Megan had cream today.
  j. Megan drinks too much coffee.
  k. It cannot be determined from the information given.

10. When she was 18 years old, the famous Mexican artist Frida Kahlo was gravely hurt in a bus accident, which left her bedbound for an entire year. It was during that year that she began to paint.

Based on the preceding statements, which of the following can we conclude is true?

  a. Frida Kahlo began to paint when she was 18 years old.
  b. Frida Kahlo never fully recovered from her accident.
  c. When she was young, Frida Kahlo did not want to be an artist.
  d. Frida Kahlo was married to another famous Mexican artist, Diego Rivera.
  e. Frida Kahlo became famous for her self-portraits.

11. Read the following statements:
The blue box is smaller than the red box, but larger than the green one.
The yellow box is the largest of all.
The red box is larger than the green box.

If the first two statements are true, the third statement is

  f. true.
  g. false.
  h. uncertain.
  j. partly true.
  k. not possible to determine from the information given.

12. Although both of Olga's parents are well-known pianists, Olga never learned to play the piano, because she has chosen the violin as her primary musical instrument. Olga is a very talented violinist and is often invited to perform at Carnegie Hall.

Based on this information, we can conclude that

  a. Olga's parents are very proud of her.
  b. Olga's parents wish she had also become a pianist.
  c. Olga plays only the violin.
  d. Olga will play at Carnegie Hall this year.
  e. Olga does not play the piano.

13. Alex and Jim are brothers. Alex is older than Jim. Jim is almost as tall as his father.

Which of the following must be true?

  f. Jim is younger than Alex.
  g. Alex is taller than his father.
  h. Alex is taller than Jim.
  j. Alex and Jim's father looks very young for his age.
  k. Alex and Jim resemble each other.

14. Juan can never fall asleep without hearing a bedtime story. Juan's parents were very busy last night, and he went to bed without a story.

Which of the following can we assume is true?

  a. Last night, Juan fell asleep right away.
  b. Juan fell asleep quickly, but woke up again.
  c. Juan's parents are usually very busy.
  d. Juan never listens to his parents.
  e. Juan could not fall asleep last night.

15. Read the following statements:
Corn flakes cost less than wheat flakes.
Oat flakes cost more than wheat flakes.
Oat flakes cost less than corn flakes.

If the first two statements are true, the third statement must be

  f. true.
  g. false.
  h. probably false.
  j. partly true
  k. not possible to determine from the information given

Reading Comprehension

This section tests your reading comprehension—your ability to understand what you read. Read each passage carefully and answer the questions that follow it. If necessary, you can reread the passage to be certain of your answers. Remember that your answers must be based only on information that is actually in the passage.

Read the following passage and answer Questions 16 through 20.

Today, bicycles are elegantly simple machines that are common around the world. Many people ride bicycles for recreation, whereas others use them as a means of transportation. The first bicycle, called a *draisienne*, was invented in Germany in 1818 by Baron Karl de Drais de Sauerbrun. Because it was made of wood, the *draisienne* wasn't very durable. It did not have pedals, and riders moved it by pushing their feet against the ground.

In 1839, Kirkpatrick Macmillan, a Scottish blacksmith, invented a much better bicycle. Macmillan's machine had tires with iron rims to keep them from getting worn down. He also used foot-operated cranks, similar to pedals, so his bicycle could be ridden at a quick pace. It didn't look much like the modern bicycle, though, because its back wheel was substantially larger than its front wheel. Although Macmillan's bicycles could be ridden easily, they were never produced in large numbers.

In 1861, Frenchman Pierre Michaux and his brother Ernest invented a bicycle with an improved crank mechanism. They called their bicycle a *vélocipède*, but most people called it a "bone shaker" because of the jarring effect of the wood and iron frame. Despite the unflattering nickname, the *vélocipède* was a hit. After a few years, the Michaux family was making hundreds of the machines annually, mostly for fun-seeking young people.

Ten years later, James Starley, an English inventor, made several innovations that revolutionized bicycle design. He made the front wheel many times larger than the back wheel, put a gear on the pedals to make the bicycle more efficient, and lightened the wheels by using wire spokes. Although this bicycle was much lighter and less tiring to ride, it was still clumsy, extremely top-heavy, and ridden mostly for entertainment.

It wasn't until 1874 that the first truly modern bicycle appeared on the scene. Invented by another Englishman, H.J. Lawson, the safety bicycle would look familiar to today's cyclists. The safety bicycle had equal-sized wheels, which made it much less prone to toppling over. Lawson also attached a chain to the pedals to drive the rear wheel. By 1893, the safety bicycle had been further improved with air-filled rubber tires, a diamond-shaped frame, and easy braking. With the improvements provided by Lawson, bicycles became extremely popular and useful for transportation. Today, they are built, used, and enjoyed all over the world.

16. Which of the following best tells what this passage is about?
   a. People should use bicycles for transportation.
   b. Bicycle manufacturers encounter many problems.
   c. a comparison of bicycles used for fun with bicycles used for transportation
   d. the history of the bicycle
   e. why bicycles have chains

17. According to the passage, which of the following is a true statement?
   f. Several people contributed to the development of the modern bicycle.
   g. Only a few *vélocipèdes* built by the Michaux family are still in existence.
   h. For most of the nineteenth century, few people rode bicycles just for fun.
   j. Bicycles with wheels of different sizes cannot be ridden easily.
   k. The best bicycle was invented in Germany.

18. The first person to use a gear system on bicycles was
   a. H.J. Lawson.
   b. Kirkpatrick Macmillan.
   c. Pierre Michaux.
   d. Ernest Michaux.
   e. James Starley.

19. Macmillan added iron rims to the tires of his bicycle to
   f. add weight to the bicycle.
   g. make the tires last longer.
   h. make the ride less bumpy.
   j. make the ride less tiring.
   k. make the bicycle more modern.

20. Which of the following statements is true?
  - **a.** James Starley, an English inventor, made several innovations that ended the progress of bicycle design.
  - **b.** James Starley, an English inventor, made several innovations that drastically changed bicycle design.
  - **c.** James Starley, an English inventor, made several innovations that quickly became outdated.
  - **d.** James Starley, an English inventor, made several innovations that gave him complete control over bicycle design.
  - **e.** James Starley, an English inventor, made several innovations that inspired other inventors to work on bicycles.

Read the following passage and answer Questions 21 through 25.

None of them knew the color of the sky. Their eyes glanced level, and were fastened upon the waves that swept toward them. These waves were of the hue of slate, save for the tops, which were foaming white, and all of the men knew the colors of the sea.

The cook squatted in the bottom as he bent to bail out the boat.

The oiler, steering with one of two oars in the boat, sometimes raised himself suddenly to keep clear of water that swirled in over the stern. It was a thin little oar and it seemed often ready to snap.

The correspondent, pulling at the other oar, watched the waves and wondered why he was there.

The injured captain, lying in the bow, was at this time buried in that profound dejection and indifference which comes, temporarily at least, to even the bravest and most enduring when, willy nilly, the firm fails, the army loses, the ship goes down.

As each slatey wall of water approached, it shut all else from the view of the men in the boat, and it was not difficult to imagine that this particular wave was the final outburst of the ocean. There was a terrible grace in the move of the waves, and they came in silence, save for the snarling of the crests.

In disjointed sentences the cook and the correspondent argued as to the difference between a lifesaving station and a house of refuge. The cook had said: "There's a house of refuge just north of the Mosquito Inlet Light, and as soon as they see us, they'll come off in their boat and pick us up."

"As soon as who sees us?" said the correspondent.

"The crew," said the cook.

"Houses of refuge don't have crews," said the correspondent.

"As I understand them, they are only places where clothes and grub are stored for the benefit of shipwrecked people. They don't carry crews."

"Oh, yes, they do," said the cook.

"No, they don't," said the correspondent.

"Well, we're not there yet, anyhow," said the oiler, in the stern.

"Bully good thing it's an on-shore wind," said the cook. "If not, where would we be? Wouldn't have a show."

In the meantime the oiler and the correspondent sat together in the same seat, and each rowed an oar. Then the oiler took both oars; then the correspondent took both oars; then the oiler, then the correspondent. The captain, rearing cautiously in the bow, after the dinghy soared on a great swell, said that he had seen the lighthouse at Mosquito Inlet.

"See it?" said the captain.

"No," said the correspondent slowly, "I didn't see anything."

"Look again," said the captain. He pointed. "It's exactly in that direction."

"Think we'll make it, captain?"

"If this wind holds and the boat don't swamp, we can't do much else," said the captain.

—Stephen Crane, from "The Open Boat" (1898)

21. Which of the following best tells what this passage is about?
   a. The survivors of a shipwreck destroy each other in an attempt to stay alive.
   b. The survivors of a shipwreck struggle to stay alive after their ship sinks.
   c. The survivors of a shipwreck find refuge in a lighthouse.
   d. The survivors of a shipwreck are picked up by a rescue boat.
   e. The survivors of a shipwreck depend on their captain to save them all.

22. Why did none of the four men know the color of the sky?
   f. They were keeping their eyes on the waves.
   g. They did not know what shade of gray the sky was.
   h. They were too tired to look at the sky.
   j. The sky was the same color as the water.
   k. The waves were too high to see the sky.

23. Why does the cook say ". . . good thing it's an onshore wind. If not, where would we be"? (An onshore wind blows from the ocean toward the land.)
   a. An offshore wind stirs up taller and more dangerous waves.
   b. An onshore wind would blow them in toward land.
   c. He is trying to cheer up his companions by telling a lie.
   d. He doesn't know in which direction the land really is.
   e. He is echoing what the captain has said.

24. Why was the captain depressed and dejected?
   f. He was sad about losing his ship.
   g. He feared that a storm was coming.
   h. He was afraid that the waves would sink the dinghy.
   j. He knew that the wind was blowing the dinghy away from land.
   k. He was weak and tired from rowing.

25. Which of the following statements best compares the correspondent's and the cook's views about a house of refuge?
   a. Both men believe that crews from a house of refuge will rescue them.
   b. The correspondent says that a house of refuge has a crew, but the cook says it does not.
   c. Both men are sure that there are no houses of refuge along the coast.
   d. The cook says that a house of refuge has a crew, but the correspondent says it does not.
   e. Both men are doubtful that they will ever find a house of refuge.

Read the following passage and answer Questions 26 through 31.

Rhode Island is the smallest state in the union, yet its history is as large as any other state's. It was the first state to declare independence from British rule prior to the Revolutionary War, and it was the first state to send soldiers to defend the nation's capital at the onset of the Civil War. The American Industrial Revolution began in Rhode Island, which led the way in producing textiles and jewelry.

The state itself actually began as two separate colonies, each established as a refuge for people with independent religious convictions. Roger Williams established a colony at the head of a river and was soon joined by many others who shared his religious views. Williams considered the colony to be blessed by God, and so he named it Providence Plantations—guided by the hand of Providence.

Soon after, another group emigrated from the Massachusetts colonies to establish a haven for their own religious views. This group was strongly influenced by a woman named Anne Hutchinson. These people established themselves on several islands situated at the mouth of the Providence River, the largest of which was called Aquidneck Island. Some thought that Aquidneck resembled the Isle of Rhodes off the coast of Greece, so the community came to be known as Rhode Island.

The early years of these little colonies were filled with trouble, as the colonists tried to establish their rights of self-governance. The colonies in Connecticut and Massachusetts tried repeatedly to establish their authority over both groups, while political battles in England used the little colonies like a tug-of-war rope.

In the end, however, the two groups—one at each end of the Providence River—joined together to become the State of Rhode Island and Providence Plantations. This is still the official name of the state today, but most of us know it simply as Rhode Island—or, more affectionately, as Little Rhody.

26. Which of the following best tells what this passage is about?
   a. Religious freedom has a long history in America.
   b. People should not allow others to tell them what to do.
   c. Rhode Island was named for the Isle of Rhodes.
   d. Rhode Island may be small, but its history is important.
   e. Rhode Island experienced trouble in the early years.

27. Which of these events happened first?
   f. the Industrial Revolution
   g. Roger Williams established Providence Plantations.
   h. the Revolutionary War
   j. Anne Hutchinson moved to Aquidneck Island.
   k. The colonies faced troubled times.

28. According to the passage, Rhode Island was settled because
   a. there was good whale hunting there.
   b. the colonies in Connecticut and Massachusetts wanted more land.
   c. people wanted independence from England.
   d. the settlers thought the islands looked like the Isle of Rhodes.
   e. people wanted greater religious freedom.

29. The group of settlers influenced by Anne Hutchinson settled
   f. in a colony of textile workers.
   g. only on Aquidneck Island.
   h. at the head of a river.
   j. in Providence Plantations.
   k. on several islands at the mouth of Providence River.

30. Rhode Island began as
   a. two colonies established as refuges for people with independent religious convictions.
   b. one of the first 13 U.S. states.
   c. two separate colonies settled by immigrants from France and England.
   d. a U.S. territory.
   e. a heavily forested wilderness.

31. What does the author mean by saying that the Rhode Island colonies were used "like a tug-of-war rope"?
  f. Rhode Island was establishing a growing trade in rope.
  g. British politicians fought for control of Rhode Island.
  h. Religious freedom created trouble in Rhode Island.
  j. Politics had people tied up in knots.
  k. Tug-of-war was a popular game among the colonists.

Read the following passage and answer Questions 32 through 38.

Greyhound racing is the sixth most popular spectator sport in the United States. Over the last decade, a growing number of greyhounds have been adopted to live out their retirement as household pets, once their racing careers are over.

Many people hesitate to adopt a retired racing greyhound because they think only very old dogs are available. Actually, even champion racers only work until they are about three-and-a-half years old. Since greyhounds usually live to between 12 and 15 years old, their retirement is much longer than their racing careers.

People worry that a greyhound will be more nervous and active than other breeds and will need large space to run. These are false impressions. Greyhounds have naturally sweet, mild dispositions, and while they love to run, they are sprinters rather than distance runners. With a few laps around a fenced-in backyard every day, they are sufficiently exercised.

Greyhounds do not make good watch-dogs, but they are very good with children, get along well with other dogs (and usually cats as well), and are affectionate and loyal. They are intelligent, well-behaved dogs, usually

housebroken in only a few days. A retired racing greyhound is a wonderful pet for almost anyone.

32. Which of the following best tells what this passage is about?
  a. how to transform a racing greyhound into a good pet
  b. why people in the dog-racing business should stop racing greyhounds
  c. why retired racing greyhounds should be adopted as pets
  d. the pros and cons of adopting a racing greyhound, presented objectively
  e. why greyhound racing is such a popular sport

33. The passage promotes the idea that a greyhound is a good pet particularly for people who
  f. do not have children.
  g. live in apartments.
  h. do not usually like dogs.
  j. already have another dog or cat.
  k. need a watchdog.

34. The passage suggests that more people would adopt retired racing greyhounds if they realized that the dogs
  a. were housebroken.
  b. were long-distance runners.
  c. were only about three-and-a-half years old.
  d. loved to be in groups of other dogs.
  e. easily learned tricks

35. According to the author, greyhounds could best be described as
  f. difficult to train.
  g. shy and retiring.
  h. nervous but passive.
  j. watchful and independent.
  k. loving and devoted.

36. Families who adopt a greyhound might expect their dog to live
   a. about three or four years.
   b. to about the age of five.
   c. to about the age of ten.
   d. up to the age of 15.
   e. a year or more.

37. One drawback of adopting a greyhound is that
   f. greyhounds are not good with children.
   g. greyhounds are old when they retire from racing.
   h. the greyhound's sensitivity makes it temperamental.
   j. greyhounds are not good watchdogs.
   k. greyhounds need a lot of room to run.

38. A retired racing greyhound available for adoption will most likely be
   a. happy to be retiring.
   b. easily housebroken.
   c. a champion.
   d. high-strung.
   e. a slow runner.

Read the following passage and answer Questions 39 through 45.

Issues of overcrowding and congestion have plagued U.S. cities since the 1940s, and the total U.S. population has increased by more than 20 million each decade since. Even 200 years ago, large population increases caused problems. From 1810 through the mid-1800s, for example, New York City's population increased, on average, 58% each decade. Amid this dramatic population increase, it was jokingly said that one could travel halfway from New York to Philadelphia quicker than one could travel the length of Broadway. This Manhattan boulevard was often in such a state of chaos that it required the forceful presence of police officers to maintain order.

The dire situation of New York's streets prompted publisher Alfred Ely Beach to search for an alternative mode of transportation. In February 1870, Beach opened a below-ground transportation system that set a precedent in subterranean travel. "Pneumatic transit," as the system was known, consisted of a 312-foot wind tunnel and a 22-passenger car propelled over the tracks by a 100-horsepower fan. While this curious solution to urban transport was not the wave of the future, it paved the way for the American subway.

However, the first subway was not built in New York. Toward the end of the nineteenth century, Boston found itself in a similar situation as New York City. Rapid population growth caused an enormous strain on traffic in the downtown area, and many commuters began to rely extensively on the street-level trolley system. Owned and operated by the West End Company, the electric-powered trolleys contended with the large number of cars and pedestrians also crowding Boston's streets. Under increasing public pressure, West End partnered with the Boston Transit Commission to fund the excavation and construction of America's first subway. This underground system, nicknamed the "T," opened on September 1, 1897.

Other American cities soon followed suit. New York opened its first subway—merely nine miles long—in October 1904. Philadelphia constructed a system combining subway lines with above-ground and elevated trolley lines, much like the one in Boston, between 1905 and 1908. Although there have been ups and downs in the popularity of subways, today most major cities, including San Francisco; Los Angeles; Miami; Baltimore; Washington, DC; and Atlanta, all have subway systems.

39. Which of the following best tells what this passage is about?
   a. the history of train travel
   b. which cities have subway systems
   c. Boston's subway system
   d. why subways came to be
   e. how subways are constructed

40. Which is NOT true of American cities?
   f. Population growth was a problem even in the nineteenth century.
   g. Population increases caused problems even 200 years ago.
   h. In the 1800s, New York City's population increased about 58% every 10 years.
   j. Population growth did not become a problem until recently.
   k. Several major cities have had problems with population growth.

41. Why were subways built?
   a. Every city needs a subway system.
   b. Cities were too congested and overcrowded.
   c. People preferred underground trains.
   d. People rely on trains to go to work.
   e. People prefer trains to other modes of transportation.

42. According to the passage, Boston built a subway system because
   f. New York City had one.
   g. the street-level trolleys could not contend with the other traffic.
   h. the West End Company partnered with the Boston Transit Commission to fund the excavation and construction.
   j. they wanted to be the first.
   k. they already had the necessary materials on hand.

43. The first subway was built in
   a. New York.
   b. Washington, DC.
   c. San Francisco.
   d. Atlanta.
   e. Boston.

44. Which of these best describes "pneumatic transit" as mentioned in the passage?
   f. electric-powered trolleys
   g. street-level passenger cars
   h. subterranean trolleys
   j. wind-powered passenger cars
   k. wind-powered surface transportation

45. According to the passage, which of the following cities probably does not have a subway system?
   a. Baltimore
   b. Atlanta
   c. Cleveland
   d. Miami
   e. San Francisco

Part 2—Math

The Math Test includes 50 multiple choice questions covering content in the following areas:

- basic math
- percentages, fractions, decimals, averages
- pre-algebra
- algebra
- substitution
- factoring
- coordinate graphing
- geometric principles
- logic
- word problems

Solve each problem and select the best answer from the choices given. It is important to keep in mind that:

- Formulas and definitions of mathematical terms and symbols are not provided.
- Diagrams other than graphs are not necessarily drawn to scale. Do not assume any relationship in a diagram unless it is specifically stated or can be figured out from the information given.
- A diagram is in one plane unless the problem specifically states that it is not.
- Graphs are drawn to scale. Unless stated otherwise, you can assume relationships according to appearance. For example, (on a graph) lines that appear to be parallel can be assumed to be parallel; likewise for concurrent lines, straight lines, collinear points, right angles, and so on.
- You must reduce all fractions to lowest terms.

46. Look at this series: J14, L11, N8, P5. . . . What term should come next?
  a. Q2
  b. Q3
  c. R2
  d. S2
  e. S3

47. $x(3x^2 + y) =$
  f. $4x^2 + xy$
  g. $4x^2 + x + y$
  h. $3x^2 + x + y$
  j. $3x^3 + 2xy$
  k. $3x^3 + xy$

48. The Laurelhurt Theatre can hold 500 people. Adult tickets cost $8.50 and student tickets cost $4.00. On an August Saturday night, the theater was completely sold out and made $2,787.50 from ticket sales. Which of the following equations could be used to determine the number of adult and student tickets that were sold that evening?
  a. $(\$4.00 + \$8.50)(500x) = \$2,787.50$
  b. $\$4.00(x) + \$8.50(x) = \$2,787.50$
  c. $(\$4.00)(x) + (\$8.50)(500 - x) = \$2,787.50$
  d. $(\$4.00 + \$8.50)(x - 500) = \$2,787.50$
  e. $(\$4.00)(x) + (\$8.50)(x) = \frac{\$2,787.50}{500}$

49. Which of the following numbers is irrational?
  f. 0
  g. $\frac{3}{4}$
  h. $\sqrt{36}$
  j. $-\frac{1}{8}$
  k. $\sqrt{2}$

50. A photograph is 5 inches wide and 8 inches long. It is enlarged so that its new length is 20 inches. Which proportion can be used to find how many inches the width of the enlarged photograph is?
  a. $\frac{x}{5} = \frac{8}{20}$
  b. $\frac{5}{20} = \frac{8}{x}$
  c. $\frac{5}{x} = \frac{20}{x}$
  d. $\frac{x}{5} = \frac{20}{8}$
  e. $\frac{5}{8} = \frac{x}{20}$

51.

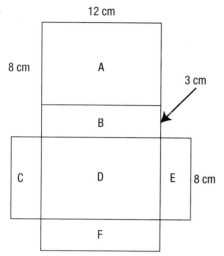

The figure shows all the faces of a rectangular prism that has been unfolded. What is the surface area in square centimeters (cm²) of the rectangular prism?

f. 312 cm²
g. 156 cm²
h. 288 cm²
j. 576 cm²
k. 31 cm²

52. A map has a scale of one inch equals five miles. If Jane and Sarah's houses are actually 30 miles apart, how far apart are the houses on the map?

a. 5 inches
b. 6 inches
c. 22 inches
d. 25 inches
e. 150 inches

53. Eight pounds of sunflower seeds, which cost 3 dollars per pound, are mixed with 18 pounds of millet, which costs 50 cents per pound. Find the cost per pound of this mixture.

f. $2.50 per pound
g. $1.97 per pound
h. $1.27 per pound
j. $1.68 per pound
k. $2.02 per pound

54. $x^2 - 4x + 4 \div (x - 2) =$

a. $x + 2$
b. $x - 2$
c. $x^2 - 2x + 2$
d. $x^2 - 3x + 2$
d. $x^2 + 2x + 2$

55. Examine (A), (B), and (C) and find the best answer.

(A): $\frac{1}{3}$ of 12
(B): 4% of 100
(C): $\frac{1}{5}$ of 10

f. (A) is less than (C).
g. (A) and (B) are equal.
h. (A) plus (B) is equal to (C).
j. (B) minus (A) is greater than (C).
k. (C) is greater than (B).

56. If $\frac{x}{3} + \frac{x}{4} = 3$, what is x?

a. $\frac{1}{12}$
b. $\frac{7}{36}$
c. $\frac{3}{4}$
d. $\frac{1}{4}$
e. $5\frac{1}{7}$

57. A bicyclist passes a farmhouse at 3:14 P.M. At 3:56 the bicyclist passes a second farmhouse. If the bicyclist is traveling at a uniform rate of 12 mph, how far apart are the farmhouses?

f. 1.2 miles
g. 1.8 miles
h. 3.6 miles
j. 8.4 miles
k. 17.1 miles

58. Annabelle is holding 6 cards: 3 spades, 1 heart, and 2 clubs. Finn will select two cards in a row out of Annabelle's hand. What is the probability that he will choose a heart followed by a spade?

   a. $\frac{1}{10}$
   b. $\frac{1}{36}$
   c. $\frac{1}{4}$
   d. $\frac{1}{3}$
   e. $\frac{1}{6}$

59. A 15-foot ladder is leaning against a wall. The base of the ladder is 9 feet from the base of the wall. How high up the wall does the top of the ladder hit the wall?

   f. 17.5 feet
   g. 12 feet
   h. 11 feet
   j. 12.8 feet
   k. 4.9 feet

60. $-6\frac{3}{4} \div 2\frac{5}{8} =$

   a. $-\frac{7}{18}$
   b. $-3\frac{5}{8}$
   c. $-2\frac{4}{7}$
   d. $-4\frac{1}{8}$
   e. $-4\frac{1}{5}$

61. The coordinates of point P and point Q are $(-2,-2)$ and $(2,3)$, respectively. How many units to the right of and above point P is point Q?

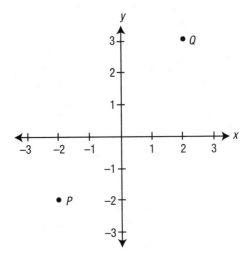

   f. one unit to the right of, and two units above, point P
   g. two units to the right of, and three units above, point P
   h. five units to the right of, and four units above, point P
   j. five units to the left of, and three units above, point P
   k. four units to the right of, and five units above, point P

62. What is the measure of the missing side of the right triangle below?

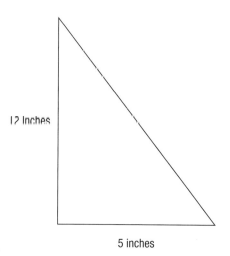

12 Inches

5 inches

    a. 7 inches

    b. 13 inches

    c. 17 inches

    d. 34 inches

    e. 60 inches

63. Which measurement is a realistic measure of the length of an average size car?

    f. $\frac{1}{4}$ kilometer

    g. 1 mile

    h. 40 cm

    j. 15 feet

    k. 12 yards

64. One night Angel spent 2 hours and 41 minutes doing homework. She spent 43 minutes on math, 37 minutes on language arts, and half an hour on science. The rest of her time was spent on social studies. How much time did she spend on social studies?

    a. 42 minutes

    b. 51 minutes

    c. 1 hour and 17 minutes

    d. 1 hour and 11 minutes

    e. 1 hour and 20 minutes

65. Examine (A), (B), and (C) and find the best answer.

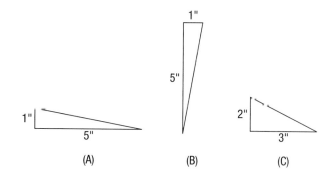

(A)   (B)   (C)

    f. The area of (A), the area of (B), and the area of (C) are all equal.

    g. The area of (A) is equal to the area of (B).

    h. The area of (C) is less than the area of (B).

    j. The area of (A) is greater than the area of (C).

    k. None of the above is true.

66. $\dfrac{6.5 \times 10^{-6}}{3.25 \times 10^{3}} =$

    a. 2×10^{9}

    b. 2×10^{-9}

    c. 2×10^{-3}

    d. 2×10^{3}

    e. 2×10^{2}

67. The length of a rectangle is equal to 4 inches more than twice the width. Three times the length plus two times the width is equal to 28 inches. What is the area of the rectangle?

    f. 8 square inches

    g. 16 square inches

    h. 20 square inches

    j. 24 square inches

    k. 28 square inches

68. If $4 + 2(3y - 2) = -10$, then $(3y - 2)$ equals

    a. -7

    b. $-\frac{5}{3}$

    c. $\frac{5}{3}$

    d. 7

    e. -3

69. Which table contains only points on this line graph?

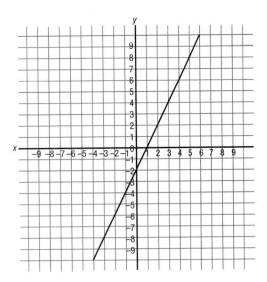

  f.

x	-2	1	3
y	-6	0	4

  g.

x	-2	0	8
y	0	1	5

  h.

x	-4	2	4
y	6	0	2

  j.

x	-1	2	4
y	-8	4	12

  k. none of the above

70. The formula for the volume of a sphere is $V = \frac{4}{3}\pi (r^3)$ where V = volume, and r = radius. If the diameter of the sphere is 10 cm, then the volume to the nearest tenth is:

    a. 4188.8 cm^3

    b. 523.6 cm^3

    c. 186.2 cm^3

    d. 62.8 cm^3

    e. 41.89 cm^3

71. Solve for x in the following equation: $\frac{1}{3}x + 3 = 8$.

    f. 33

    g. 15

    h. 11

    j. 8

    k. 3

72. Evaluate the expression $\frac{2(a-b)}{3}$ if $a = -2$ and $b = 4$.

    a. -24

    b. -4

    c. $\frac{8}{3}$

    d. 8

    e. 24

73. The table shows the number of cases of a particular kind of floor tile in four warehouses for the Home Fixers store.

Warehouse	Cases of Tiles
1	$1\frac{1}{16}$
2	$1\frac{5}{12}$
3	$1\frac{1}{4}$
4	$1\frac{2}{3}$

Hatim needs to fill an order for 3 cases of these floor tiles. From which two warehouses should he request tiles to fill this order?
  f. 1 and 2
  g. 1 and 4
  h. 2 and 3
  j. 3 and 4
  k. 2 and 4

74. What is the value of 8.9×10^7?
  a. 0.000000089
  b. 0.000000089
  c. 8,900,000
  d. 89,000,000
  e. 890,000,000

75. Bridget wants to hang a garland of silk flowers all around the ceiling of a square room. Each side of the room is 9 feet long; the garlands are only available in 15 foot lengths. How many garlands will she need to buy?
  f. 2 garlands
  g. 3 garlands
  h. 4 garlands
  j. 5 garlands
  k. 6 garlands

Use the following chart to answer Questions 81 and 82.

The chart lists the number of members present at the monthly meetings for the Environmental Protection Club.

MONTH	# OF MEMBERS
September	54
October	61
November	70
December	75

76. What was the average monthly attendance over the course of all the months listed?
  a. 61
  b. 56
  c. 63
  d. 65
  e. 71

77. If the data presented in the table were plotted as a bar graph, which of the following would best represent the data most accurately?

f.

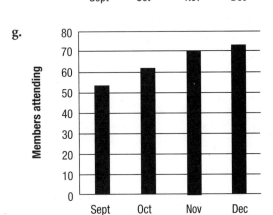

g.

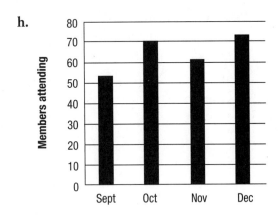

h.

j.

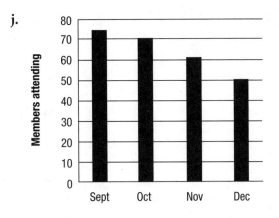

k. None of these graphs accurately displays the data.

78. Two times a number is the result when 7 times a number is taken away from 99. What is the number?
   a. 85
   b. 11
   c. 37
   d. 46
   e. 9

79. At the grocery store, Jacob purchases two pounds of chicken for $5.49 per pound, three pounds of apples for $1.89 per pound, two pounds of grapes for $2.99 per pound, and a loaf of bread for $2.59. If Jacob gives the cashier $30, how much change will he get?
   f. $7.04
   g. $4.78
   h. $5.22
   j. $8.56
   k. Jacob did not have enough money for this purchase.

80. One side of a regular octagon has a length of 4 cm. What is the perimeter of the octagon?
  a. 8 cm
  b. 24 cm
  c. 36 cm
  d. 32 cm
  e. 38 cm

81. The Landmark Theater contains 3 sections of 19 rows on the main level. Each row contains 12 seats. The balcony has 2 sections of 8 rows, with 8 seats in each row. How many people can the theater seat?
  f. 292
  g. 486
  h. 684
  j. 748
  k. 812

82. Which point on the number line best represents $\sqrt{11}$?

  a. J
  b. K
  c. L
  d. M
  e. $\sqrt{11}$ is not represented on this number line

83. The angles of a quadrilateral are in the ratio 1:2:2:4. What is the measure, in degrees, of the largest angle?
  f. 40°
  g. 50°
  h. 60°
  j. 80°
  k. 160°

84. Which figure best represents a triangle with sides a, b, and c in which the relationship $a^2 + b^2 = c^2$ is always true?

  a.

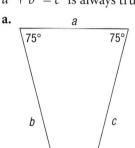

  b.

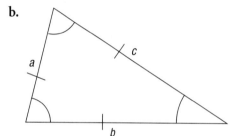

  c.

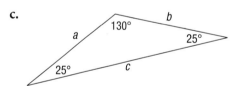

  d.

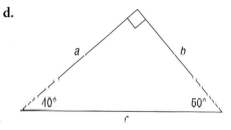

  e. The relationship $a^2 + b^2 = c^2$ is true for all triangles.

85. The White-Bright Toothbrush Company hired 30 new employees. This hiring increased the company's total workforce by 5%. How many employees now work at White-Bright?
   f. 530
   g. 600
   h. 605
   j. 630
   k. 620

Use the following table to answer Questions 86 through 88.

The table lists the size of building lots in the Orange Grove subdivision and the people who are planning to build on those lots. For each lot, installation of utilities costs $12,516. The city charges impact fees of $3,879 per lot. There are also development fees of 16.15 cents per square foot of land.

Lot	Area (sq.ft.)	Builder
A	8,023	Ira Taylor
B	6,699	Alexis Funes
C	9,004	Ira Taylor
D	8,900	Mark Smith
E	8,301	Alexis Funes
F	8,269	Ira Taylor
G	6,774	Ira Taylor

86. How much will Ms. Funes have to pay for installation of utilities and impact fees to develop her lots?
   a. $37,635
   b. $32,790
   c. $24,862
   d. $16,395
   e. $4,845

87. How much land does Mr. Taylor own in the Orange Grove subdivision?
   f. 23,066 sq. ft
   g. 26,842 sq. ft
   h. 29,765 sq. ft
   j. 31,950 sq. ft
   k. 32,070 sq. ft

88. How much will Mr. Smith pay in development fees for his lot?
   a. $1,157.00
   b. $1,437.35
   c. $96,540.00
   d. $143,735.00
   e. $274,550.00

89. A taxi ride costs $2.50 for the first mile and $0.50 for every mile after that. Dana gave the cab driver $20. This included a tip of $3.

If she received $4.50 in change, how many miles did she travel in the cab?
   f. 29
   g. 30
   h. 20
   j. 21
   k. 12.5

90. During a trip to Yosemite National Park, Jason observed a climber high on a rock wall. Jason was standing 125 meters from the vertical rock face. Use the diagram below to find how high the climber was above the ground.

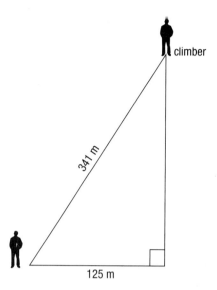

a. 317.3 m
b. 329.2 m
c. 363.2 m
d. 423.7 m
e. 466.1 m

91. Choose the statement that is logically equivalent to "If I earn an A on the test, then I will be pleased."
   f. If I am pleased, then I will earn an A on the test.
   g. If I do not earn an A on the test, then I will not be pleased.
   h. If I am not pleased, then I did not earn an A on the test.
   j. If I earn an A, then I am not pleased.
   k. If I am not pleased, then I will not earn an A on the test.

92. What is the probability that the spinner will land on an even number?

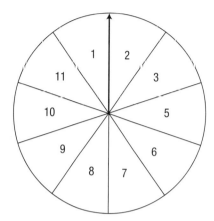

   a. $\frac{2}{3}$
   b. $\frac{4}{11}$
   c. $\frac{3}{5}$
   d. $\frac{1}{2}$
   e. $\frac{2}{5}$

93. Which of the following numbers is NOT a factor of 36?
   f. 6
   g. 1
   h. 12
   j. 2
   k. 13

94. What is the prime factorization of 48?
   a. $2 \times 4 \times 6$
   b. 2×24
   c. $2 \times 3 \times 7$
   d. $2 \times 3 \times 8$
   e. $2 \times 2 \times 2 \times 2 \times 3$

95. If *n* is the position of a number in this sequence, which expression identifies this pattern?

Position	1st	2nd	3rd	4th	...	*n*th
Value of the Term	2	6	14	30	...	

    f. $2^n + 1$
    g. $2^n + 2$
    h. $2^{(n+1)} + 1$
    j. $2^{(n+1)} + 2$
    k. $2^{(n+1)} - 2$

Answers

Paragraph 1 (U, R, S, Q, T)

Note the logical order of the action in this story. After Apollo fell in love with the nymph (opening sentence), he confessed his love (**U**). After Daphne began to run away, Apollo chased her (**R**). After she prayed to her father for help (**S**), he transformed her into a tree (**Q**). Sentence **T** is the resolution to the action.

Paragraph 2 (S, R, U, T, Q)

Pay attention to the chronological order of the actions described. Sentence **S** is most logically placed just after the opening sentence, leading to the events that follow. Sentence **R** introduces the events that happened in Kropp's adult life. **U** describes a bicycle journey *to* Mt. Everest, **T** is about climbing the mountain, and **Q** is about a bicycle journey home *after* climbing Mt. Everest.

Paragraph 3 (R, S, U, Q, T)

Sentence **R** logically follows the introductory sentence, because it explains why answering the telephone is so important. Sentence **S** refers back to **R** as the reason for needing impeccable manners. The actions mentioned then take place in an obvious sequence: you answer the phone when it rings, **U**, give a short, friendly greeting, **Q**, and take messages, **T**.

Paragraph 4 (S, R, U, Q, T)

The sentence given introduces the idea of how yoga requires a lot of effort. Sentence **S** refers back to that effort. It would not work to put **S** at the end of the paragraph because the phrase "Even so" would be meaningless there. Sentence **R** mentions a specific example of a sitting pose that stretches and strengthens muscles. Sentence **U** describes that sitting pose. Sentence **Q** mentions a standing pose that is "even more difficult than *staff pose*," so this sentence could not go before **R** or **U**. Sentence **T** describes that standing pose.

Paragraph 5 (T, U, R, Q, S)

The given sentence states the main idea of the paragraph: you should plan your holiday party early. Sentence **T** most logically follows the introductory sentence because it states one reason for planning early. **T** mentions popular holidays and **U** continues the train of thought about planning ahead for a party during "one of these holidays." Note that the word *also* in choices **R** and **S** indicates that those sentences follow some other thought. Sentence **R** first mentions invitations. Sentence **Q** gives says what to include on the invitations, and sentence **S** gives what you might *also* want to include on the invitations. That gives you a clear sentence order for **R**, **Q**, and **S**.

    6. c. If Randy is two months older than Greg, then Ned is three months older than Greg and one month older than Randy. Kent is younger than both Randy and Ned. Ned is the oldest.

    7. j. Statements II and III are the only true statements. Since all dogs like to run, then the ones who like to swim must like to run. Since all dogs like to run, then the ones who look like their masters must like to run. There is no support for statement I.

8. b. Based on the statements given, we can conclude that Consuela spent less time driving than Mark (choice **b**), because we are told that she spent 36 hours behind the wheel, while Mark spent 48 hours. We do not know (based only on the statements) what kind of drivers Mark and Consuela are (choice **a**) or who is a more experienced driver (choice **d**). Just because Mark spent more time behind the wheel does not mean that he drove below the speed limit (choice **c**) or that Consuela doesn't like to drive (choice **e**).

9. h. We are told that Megan never drinks black coffee, but we are *not* told why. Therefore, we cannot conclude that she doesn't like coffee black. She may or may not have put sugar in her three cups, but we know that she had cream because she always uses cream.

10. a. From the information provided to us, we know that Frida Kahlo became bedbound at the age of 18 as a result of a bus accident, and it was during that time that she began to paint. Therefore, we can conclude that Frida Kahlo began to paint when she was 18 years old (choice **a**). Although all the other statements may be true, we cannot make these conclusions based on the information provided to us.

11. f. One way to approach this problem is to translate the information into simple "real world" terms. Give the boxes sizes that fit the statements:

blue (smaller than red; larger than green), 10 inches

red (larger than blue), 12 inches

green (smaller than blue), 8 inches

yellow (largest), 15 inches

Then reorder this list by size:

yellow

red

blue

green

It instantly becomes apparent that the red box is larger than the green one, making the third statement true.

12. e. Although we know many facts about Olga, we cannot assume from the information given to us whether Olga's parents are proud of her (choice **a**), or whether they wish she had also become a pianist (choice **b**). Because we are told that violin is Olga's *primary* instrument, we can assume that she may also play other instruments. Although she often plays at Carnegie Hall, we cannot assume that she will play there this year (choice **d**). The only information we know for certain is that Olga does not play the piano (choice **e**).

13. f. Based on the information available to us, the only answer choice we can conclude is correct is that Jim is younger than Alex (choice **f**), since we are told that Alex is older than Jim. Since there is no mention of how tall Alex and Jim's father is or what any of them look like, we cannot make any assumptions for choices **g**, **h**, **j**, and **k**.

14. e. Based on the given information, the only thing we know for certain is that Juan could not fall asleep last night (choice **e**), because we are told he can never fall asleep without hearing a bedtime story. We can also assume that Juan could not have fallen asleep right away last night (choices **a** and **b**). There is no evidence to verify information in choices **c** and **d**.

15. g. Real world terms can help with this problem. For example, you can give dollar amounts to each statement:

    wheat flakes, $2.50
    corn flakes (less than wheat), $2.00
    oat flakes (more than wheat), $2.75

The statement that "oat flakes (at $2.75) cost less than corn flakes (at $2.00)" is clearly false, and the answer is choice **g**.

16. d. The passage gives the history of the bicycle. Choice **a** is incorrect because few opinions are included in the passage. There is no support for choices **b**, **c**, and **e**.

17. f. This is the best choice because each paragraph of the passage describes an inventor whose machine was a step toward the modern bicycle. There is no evidence to support choice **g**. Choices **h**, **j**, and **k** are incorrect because they make statements that, according to the passage, are untrue.

18. e. The fourth paragraph states that *James Starley* added a gear to the pedals.

19. g. This information is clearly stated in the second paragraph. The iron rims kept the tires from getting worn down, and, therefore, the tires lasted longer. Choice **f** is incorrect because although the iron rims probably did make the machine heavier, that was not Macmillan's goal. Choice **h** is incorrect because no information is given about whether iron-rimmed or wooden tires moved more smoothly. There is no support for choices **j** or **k**.

20. b. Based on the paragraph, this is the only possible choice. Starley *revolutionized* the bicycle; that is, he made many innovative changes. Based on the context, the other choices make no sense.

21. b. Although the other choices are things that might happen, the passage gives no evidence of any those choices having already happened.

22. f. The first paragraph implies that the reason the men did not know the color of the sky was that they were too busy watching the waves. This was the only way to avoid being hit by a wave that would capsize their frail lifeboat. Choices **g**, **h**, **j**, and **k** might seem plausible, but they are not supported in the passage.

23. b. Choice **a** is incorrect because an offshore wind comes from land and would blow the boat farther out to sea. There is no evidence in the text to support choices **c**, **d**, or **e**. All the sailors know the direction in which the land lies, and a little bit later, the captain does see signs of land.

24. f. The text makes it clear that the captain is dejected because his ship has sunk. Choice **g** is incorrect because nothing in the text suggests that a storm is coming. Choice **h** is incorrect because no one, least of all the captain, has given up hope of surviving the waves. Choice **j** is incorrect because the wind is blowing toward shore, not away from it. Choice **k** is not correct since the correspondent and the oiler, not the captain, are rowing. The captain is in charge of the dinghy and its small crew.

25. d. In the conversation between the correspondent and the cook, it is the cook who says that a house of refuge has a crew. The correspondent says it doesn't. So choice **d** is the correct answer. The passage does not support any of the other answers.

26. d. The passage does address religious freedom (choice **a**), but it does not trace its history in America. The main idea of the passage is that Rhode Island has an important history, despite its small size. Choice **b** is not supported by the passage, and the remaining choices are details from the passage.

27. g. Roger Williams was the first to colonize the Providence area. Anne Hutchinson (choice **j**) and others followed later. The troubles (choice **k**) followed colonization, and this all occurred long before the Revolutionary War (choice **h**) and the Industrial Revolution (choice **f**).

28. e. Each of the choices is actually true in Rhode Island's history, but the reason that it was originally settled was for religious freedom (choice **e**), as stated in the passage.

29. k. The passages states, "These people established themselves on several islands situated at the mouth of the Providence River." The other choices refer to other settlers or individuals.

30. a. The passage states, "The state itself actually began as two separate colonies, each established as a refuge for people with independent religious convictions." The passage does not support choices **c** and **d**. Choice **e** may be true, but is not mentioned in the passage.

31. g. The author uses a tug-of-war rope to picture two political groups fighting, with Rhode Island caught in the middle. Each group was trying to pull Rhode Island away from the other group. The other choices are not supported by the passage.

32. c. The tone of the passage is enthusiastic in its recommendation of the greyhound as a pet and thereby encourages people to adopt one. It does not give advice on transforming a greyhound (choice **a**). The passage does not address the dog-racing audience (choice **b**) or discuss the popularity of the sport (choice **e**). The author is not objective (choice **d**) because he or she only presents the pros about adopting greyhounds.

33. j. See the last paragraph. The passage does not mention choice **g** or choice **h**. Choices **f** and **k** are clearly wrong; the passage states the opposite.

34. c. The first two sentences in the second paragraph support this answer. Choice **a** is incorrect because the passage states that greyhounds are easily housebroken, not that they are already housebroken. Choice **b** is contradicted in the passage. There is no support for choices **d** or **e**.

35. k. In the last paragraph, greyhounds are described as *affectionate and loyal*, which is the same as *loving and devoted*. The other choices are incorrect, according to the information given.

36. d. This detail is stated directly in the second paragraph.

37. j. See the last paragraph. Choices **f**, **g**, **h**, and **k** are contradicted in the passage.

38. b. See the end of the next-to-last sentence in the passage. Choices **a**, **c**, **d**, and **e** are not to be found in the passage.

39. d. The passage states that overcrowding and population increases were the reasons subways were built. Choice **a** is not accurate because trains are not necessarily only subways. Choices **b** and **c** are true, but they are only part of the passage's topic. Choice **e** is not addressed in the passage.

40. j. Note that the question asks for the answer that is not true. According to the passage, choice **j** is false and all the other choices are true.

41. b. The first paragraph describes the overcrowding conditions that led to the construction of subways. Choices **a** and **c** are not necessarily true. Choices **d** and **e** are not discussed in the passage.

42. g. The third paragraph describes the crowding of Boston's streets. Choice **f** is not true. Choice **h** does not describe why the system was built. We do not know whether choice **j** is true, and the passage does not mention materials (choice **k**).

43. e. The passage clearly states that the first subway was not built in New York (choice **a**), but in Boston. The other cities listed have subways today, but none of them was the first.

44. j. Pneumatic transit was subterranean, but more importantly, it was powered by wind. Therefore the best answer is **j**.

45. c. Cleveland is the only city in this list that the passage does not say has a subway system.

46. c. R2. The letter part of each pairing is skipping ahead by one and the number part of each pairing is decreasing by 3.

47. k. $3x^3 + xy$. The distributive property states that x will be multiplied by both terms inside the parentheses—exponents are added when like bases are multiplied.

48. c. $(\$4.00)(x) + (\$8.50)(500 - x) = \$2,787.50$. Allow the number of student tickets sold to be x. Since all 500 seats were sold, then the number of adult tickets sold would be $(500 - x)$. Multiply the expression representing the number of the student tickets sold by the student ticket price: $(\$4.00)(x)$; and do the same for the adult tickets: $(\$8.50)(500 - x)$. Their sum will be the total amount of money the movie theater made that night.

49. k. $\sqrt{2} \cdot \sqrt{2}$ is irrational because it does not terminate.

50. e. $\frac{5}{8} = \frac{x}{20}$. In this proportion, "width" is over "length" in both fractions.

51. f. 312 cm². Area of a rectangle = (base x width) and each face can be calculated individually: A: 8 cm × 12 cm = 96 cm², B: 3 cm × 12 cm = 36 cm², C: 8 cm × 3 cm = 24 cm², D: 8 cm × 12 cm = 96 cm², E: 8 cm × 3 cm = 24 cm², F: 3 cm × 12 cm = 36 cm² and the sum of these areas is 312.

52. b. 6 inches. Set up a proportion with "inches ÷ miles": $\frac{1}{5} = \frac{x}{30}$ and solving this you get $x = 6$.

53. h. $1.27 per pound. (8 lb)($3) + (18 lb)($0.50) = $33.00 for 26 pounds of mixture. To find the per pound cost, divide $33 by 26 pounds.

54. b. $x - 2$. Starting with $x^2 - 4x + 4 \div (x - 2)$, you should see that the numerator can be factored using the rule for perfect squares: $x^2 - 4x + 4 = (x - 2)(x - 2)$. Then one of the $(x - 2)$ terms in the numerator will cancel out with the $(x - 2)$ in the denominator. All that will remain will be $(x - 2)$.

55. g. **(A) and (B) are equal.** (A): $\frac{1}{3}$ of 12 = 4, (B): 4% of 100 = 4, (C): $\frac{1}{5}$ of 10 = 2

56. e. $5\frac{1}{7}$. Multiply all terms by 12 to get rid of the fractions: $12(\frac{x}{3} + \frac{x}{4}) = 12(3)$ results in $4x + 3x = 36$ and then $7x = 36$ so then $x = \frac{36}{7} = 5\frac{1}{7}$.

57. j. **8.4 miles.** Remember the formula distance = rate × time where time is expressed in hours. 42 minutes is then $\frac{42}{60}$ hours, since there are 60 minutes in an hour. Therefore, distance = 12 mph × $\frac{42}{60}$ = 8.4 miles.

58. a. $\frac{1}{10}$. The probability that a single desirable event will happen is always $\frac{\text{number of desirable events}}{\text{total number of events}}$. In this case the probability that a heart will be pulled is $\frac{1}{6}$. Then there are 3 spades left out of the remaining 5 cards so the next probability is $\frac{3}{5}$. The probability that two events will happen in a row is the product of each individual probability: $\frac{1}{6} \times \frac{3}{5} = \frac{3}{30} = \frac{1}{10}$.

59. g. 12 feet. This example makes a right triangle, so use the Pythagorean theorem: $a^2 + b^2 = c^2$; $9^2 + b^2 = 15^2$; $81 + b^2 = 225$; $b^2 = 144$.

60. c. $-2\frac{4}{7}$. First turn each mixed fraction into an improper fraction: $-\frac{27}{4} \div \frac{21}{8}$. Then turn the equation into multiplication and use the reciprocal of the second fraction: $-\frac{27}{4} \times \frac{8}{21} = -\frac{216}{8} = -\frac{18}{7}$.

61. k. four units to the right of, and five units above, point P. Moving right from −2 to 2 takes 4 units, and moving up from −2 to 3 takes 5 units.

62. b. 13 inches. Use the Pythagorean theorem for this right triangle: $a^2 + b^2 = c^2$; $12^2 + 5^2 = c^2$; $144 + 25 = c^2$; $c^2 = 169$, $c = 13$.

63. j. 15 feet. This is the only reasonable measure for an average car since 40 cm is less than 2 feet and 15 yards is 45 feet.

64. b. 51 minutes. Angel spent (2 hrs) × (60 min) + 41 min = 161 minutes doing homework. $161 − 43 − 37 − 30 = 51$ minutes.

65. g. The area of (A) is equal to the area of (B). Area of a triangle = $\frac{1}{2}$(base × height). Area of A: 2.5 in²; Area of B: 2.5 in²; Area of C: 3 in².

66. b. 2×10^{-9}. To divide scientific notation, divide 6.5 by 3.25 and then use exponent rules to divide the bases of 10.

67. g. 16 square inches. Let width = w and length = $2w + 4$. 3(length) + 2(width) = 28; next is $3(2w + 4) + 2(w) = 28$; next is $6w + 12 + 2w = 28$; $8w = 16$ so $w = 2$ and $l = 8$. Area = (length)(width) = 16.

68. a. −7. Given $4 + 2(3y − 2) = −10$, then $2(3y − 2) = −14$, and $(3y − 2) = −7$.

69. f.

x	−2	1	3
y	−6	0	4

When graphing coordinate pairs, start at the origin (0,0) and move horizontally first for your x-coordinate, and vertically next for your y-coordinate. In choice **g**, (−2,0) is not on the line; in choice **h**, (−4,6) is not on the line; in choice **j**, (−1,8) is not on the line.

70. b. 523.6 cm³. The radius is half the diameter, so radius = 5 cm. $V = \frac{4}{3}\pi (5^3) = V = \frac{4}{3}(3.14)(5^3) = 523.6$

71. g. 15. $\frac{1}{3}x + 3 − 3 = 8 − 3$; so $\frac{1}{3}x = 5$ and $x = 15$.

72. b. −4. $\frac{2(a − b)}{3} = \frac{2(−2 − 4)}{3} = \frac{2(−6)}{3} = \frac{−12}{3} = −4$.

73. k. 2 and 4. $1\frac{5}{12} + 1\frac{2}{3} = 1\frac{5}{12} + 1\frac{8}{12} = 2\frac{13}{12} = 3\frac{1}{12}$.

74. d. 89,000,000. Move the decimal 7 times to the right, which will result in 6 zeros after the 9.

75. g. 3 garlands. The perimeter of the room will be 9 × 4 = 36 feet. 2 garlands will only be 30 feet long, so 3 are needed.

76. d. 65. To find the average of 4 numbers, find their sum and divide by 4: $\frac{54 + 61 + 70 + 75}{4} = \frac{260}{4} = 65$

77. g. This bar graph represents the steady increase of the number of members.

78. b. 11. Begin with $2x = 99 − 7x$. Then $9x = 99$ and $x = 11$.

79. g. $4.78. ($5.49)(2lbs) + ($1.89)(3lbs) + ($2.99)(2lbs) + $2.59 = $25.22, which is the total. $30.00 − $25.22 = $4.78.

80. d. 32 cm. Octagons have 8 sides so the perimeter will be 8 × 4 = 32 cm.

81. k. 812. Main level = 3 × 19 × 12 = 684 seats. Balcony = 2 × 8 × 8 = 128. 684 + 128 = 812.

82. b. Point K. $\sqrt{9} < \sqrt{11} < \sqrt{16}$ which means $\sqrt{11}$ must be between 3 and 4, since $\sqrt{9} = 3$ and $\sqrt{16} = 4$. Therefore, **b** is the best choice.

83. k. 160°. The angles of a quadrilateral sum to 360°. Set up the equation $1x + 2x + 2x + 4x = 360°$. So $9x = 360°$ and $x = 40°$. The largest angle is $4x$ so it will measure 160°.

84. d. The relationship $a^2 + b^2 = c^2$ is always true for right triangles only.

85. j. 630. 30 new employees = (5%)(original # of employees); so $30 = .05y$; and $y = 600$. There were originally 600 employees, so now there are 630.

86. b. $32,790. Ms. Funes will have to pay the utility cost and the impact fee twice each, one for each property. She will have to pay $3,879(2) + 12,516(2) = 32,790.

87. k. 32,070 sq. ft. $8,023 + 9,004 + 8,269 + 6,774 = 32,070$.

88. b. $1,437.35. $8,900(\$0.1615) = \$1,437.35$.

89. j. 21 miles. The taxi's fee can be represented with the equation: $\$2.50 + (m - 1)(\$0.50)$. We use $(m - 1)$ since the first mile is included in the initial $2.50 fee. Dana gave the cab driver $20 including a tip of $3 and got back $4.50 in change, her fare must have been $20 - $3 - $4.50 = $12.50. Therefore, $\$12.50 = \$2.50 + (m - 1)(\$0.50)$; which simplifies to $\$10 = \$0.50m - \$0.50$; so $\$0.50m = \$10.50$ and $m = 21$ miles.

90. a. 317.3 m. Use the Pythagorean theorem for this right triangle: $a^2 + b^2 = c^2$; $125^2 + b^2 = 341^2$; $15,625 + b^2 = 116,281$; $b^2 = 100,656$; $b = 317.26$m.

91. h. If I am not pleased, then I did not earn an A on the test. The contrapositive of a statement is always equivalent to the original statement. The contrapositive is made by switching the order of the two parts and then negating each part.

92. e. $\frac{2}{5}$. The probability that a single desirable event will happen is always $\frac{\text{number of desirable events}}{\text{total number of events}}$. In this case the total number of events is 10 since the spinner is divided into 10 parts. The total number of desirable events is 4 since there are 4 even numbers. Therefore the probability is $\frac{4}{10} = \frac{2}{5}$.

93. k. 13. A factor is a number that divides another number. 13 does not divide 36.

94. e. $2 \times 2 \times 2 \times 2 \times 3$. The prime factorization of $48 = 24 \times 2 = 6 \times 4 \times 2 = (3 \times 2) \times (2 \times 2) \times 2 = 2 \times 2 \times 2 \times 2 \times 3$.

95. k. $2^{(n+1)} - 2$. Sub in values for n to see which choice works: when $n = 1$, $2^{(1+1)} - 2 = 2$; when $n = 2$, $2^{(2+1)} - 2 = 6$; when $n = 3$, $2^{(3+1)} - 2 = 14$; when $n = 4$, $2^{(4+1)} - 2 = 30$.

PRACTICE TEST 2

C H A P T E R

The *SHSAT Power Practice* tests will help you prepare for the high-stakes exams given to students applying for New York City's specialized high schools. Each practice test consists of sample questions like
those you will find on the official SHSAT.

The 45-question verbal section and 50-question math section were developed by education experts. These
tests will show you how much you know and what kinds of problems you still need to study. Mastering these
practice tests will allow you to reach your highest potential on the real SHSAT.

PART I VERBAL

Scrambled Paragraphs

Paragraph 1

q r s t u
q r s t u
q r s t u
q r s t u
q r s t u

Paragraph 2

q r s t u
q r s t u
q r s t u
q r s t u
q r s t u

Paragraph 3

q r s t u
q r s t u
q r s t u
q r s t u
q r s t u

Paragraph 4

q r s t u
q r s t u
q r s t u
q r s t u
q r s t u

Paragraph 5

q r s t u
q r s t u
q r s t u
q r s t u
q r s t u

Logical Reasoning

6. a b c d e
7. f g h j k
8. a b c d e
9. f g h j k
10. a b c d e
11. f g h j k
12. a b c d e
13. f g h j k
14. a b c d e
15. f g h j k

Reading

16. a b c d e
17. f g h j k
18. a b c d e
19. f g h j k
20. a b c d e
21. f g h j k
22. a b c d e
23. f g h j k
24. a b c d e
25. f g h j k
26. a b c d e
27. f g h j k
28. a b c d e
29. f g h j k
30. a b c d e

31. f g h j k
32. a b c d e
33. f g h j k
34. a b c d e
35. f g h j k
36. a b c d e
37. f g h j k
38. a b c d e
39. f g h j k
40. a b c d e
41. f g h j k
42. a b c d e
43. f g h j k
44. a b c d e
45. f g h j k

PART II MATHEMATICS

46. a b c d e
47. f g h j k
48. a b c d e
49. f g h j k
50. a b c d e
51. f g h j k
52. a b c d e
53. f g h j k
54. a b c d e
55. f g h j k
56. a b c d e
57. f g h j k
58. a b c d e
59. f g h j k
60. a b c d e
61. f g h j k
62. a b c d e

63. f g h j k
64. a b c d e
65. f g h j k
66. a b c d e
67. f g h j k
68. a b c d e
69. f g h j k
70. a b c d e
71. f g h j k
72. a b c d e
73. f g h j k
74. a b c d e
75. f g h j k
76. a b c d e
77. f g h j k
78. a b c d e
79. f g h j k

80. a b c d e
81. f g h j k
82. a b c d e
83. f g h j k
84. a b c d e
85. f g h j k
86. a b c d e
87. f g h j k
88. a b c d e
89. f g h j k
90. a b c d e
91. f g h j k
92. a b c d e
93. f g h j k
94. a b c d e
95. f g h j k

Part 1—Verbal

The Verbal Test includes 45 questions in these three sections:

- Scrambled Paragraphs, 5 paragraphs (each counts double)
- Logical Reasoning, 10 questions, numbered 6–15
- Reading, 30 questions, numbered 16–45

Scrambled Paragraphs

This section tests your ability to organize a paragraph well. There are five paragraphs, presented in scrambled order. Your job is to put them in the best order to make a clear, coherent paragraph. Each correct answer counts double; these five paragraphs are worth 10 points out of the 50-point verbal test.

The first sentence in each paragraph is given. The remaining five sentences are listed in random order. Read each group of sentences carefully, and then decide on the best arrangement for them. Use the blanks at the left of each sentence to number these sentences from 1 to 5, showing the order they should be in.

Paragraph 1
There are many myths about bats that need to be dispelled if we are to learn to appreciate these fascinating creatures.

_____ Q. The anatomy of their arms and hand bones is very much like our own.

_____ R. For example, bats are not "flying mice," as they have sometimes been described.

_____ S. Actually, bats are more closely related to primates (including humans) than they are to rodents.

_____ T. It is true that they look a little like winged mice, so describing them that way was an easy mistake to make.

_____ U. However, scientists say that bats are not related to mice at all.

Paragraph 2
Some authors live dangerous lives and write about their own experiences.

_____ Q. Emily Brontë is an example of a writer who did not seek danger.

_____ R. It is also true that F. Scott Fitzgerald wrote about a fast-paced life that eventually destroyed him.

_____ S. It is well-known that Ernest Hemingway went to war to gather material for his stories.

_____ T. However, it is not necessary for a writer to endanger his or her life in order to have something to write about.

_____ U. Brontë seldom ventured outside her father's tiny country rectory, yet she wrote *Wuthering Heights*, one of the greatest works in the English language.

Paragraph 3
Repotting a plant can be quite successful when done correctly.

_____ Q. Then choose a new pot that is one or one or two inches larger in diameter than the previous pot, and put potting soil in the bottom.

_____ R. Gently lower the plant into the new pot and fill in the sides with potting soil.

_____ S. First, hold the plant and its pot upside down.

_____ T. Once you have removed the old pot, gently massage the plant's roots, loosening them from a tight ball.

_____ U. Carefully tap the sides of the pot until you can remove it from the plant.

Paragraph 4

Memo to all staff members: This Saturday and Sunday, March 8 and 9, Under Your Feet carpet company will be installing new carpets throughout the building.

_____ Q. All office areas that are not currently carpeted will also be carpeted.

_____ R. However, offices with green carpeting will not be recarpeted, since the green carpet was installed only one year ago.

_____ S. All office areas that currently have gold carpeting will get new carpeting.

_____ T. In both of those office areas, the new carpet will be dark blue.

_____ U. Here are the guidelines for the new installation:

Paragraph 5

The word _China_ comes from the Ch'in Dynasty.

_____ Q. Then King Cheng named himself Shih Huang-ti, or First Emperor of the Ch'in Dynasty.

_____ R. Emperor Cheng next established a system of laws and a common written language.

_____ S. He also built roads and canals to the capital.

_____ T. First, the Ch'in Dynasty unified the country by conquering warring feudal lords.

_____ U. The new emperor consolidated his empire by abolishing feudal rule and creating a centralized monarchy.

Logical Reasoning

The questions in this section test your ability to reason well, that is, to figure out what the facts you know can or can't possibly mean. Read the statements carefully, then choose the best answer based _only_ on the information given. Note carefully the words used in each question. For example, one thing can be lar_ger_ than another without being the lar_gest_ in the group. In answering some of these questions, it may be useful to draw a rough diagram or make a list that gives real-world values to the information.

6. Islands are surrounded by water. Maui is an island. Maui was formed by a volcano.

    Based only on the information provided, which of the following statements must be true?

    I. Maui is surrounded by water.
    II. All islands are formed by volcanoes.
    III. All volcanoes are on islands.

    a. I only
    b. III only
    c. I and II only
    d. I and III only
    e. I, II, and III

Read the following, then answer Questions 7 and 8.

The code below has the following rules:

1. Each letter in the code represents the same word in all three sentences.
2. Each word is represented by only one letter.
3. The position of a letter in any of the code lines is _never_ the same as the position of the word it represents in the sentence.

S K B G L means
"Where is the train station?"

S B L D G means
"Where is the bus station?"

D R S B G means
"Where is the bus depot?"

7. Which word is represented by the letter **D**?
 f. where
 g. station
 h. bus
 j. either bus or train, but it cannot be determined which one
 k. either station or depot, but it cannot be determined which one

8. What letter represents the word "is"?
 a. S
 b. L
 c. B
 d. G
 e. It cannot be determined from the information given.

9. At the baseball game, Henry is sitting in seat 253. Marla is sitting to the right of Henry in seat 254. In the seat to the left of Henry is George. Inez is sitting in the seat to the left of George.

   Which seat is Inez sitting in?

 f. 251
 g. 254
 h. 255
 j. 256
 k. It cannot be determined from the information given.

10. Last week, Lia celebrated her birthday. She received many presents, flowers, and a large box of chocolates, which she has not yet opened.

   Which of the following can we conclude is true?

 a. Lia has not yet opened the chocolates because she plans to share them with her friends.
 b. Lia received a box of chocolates for her birthday.
 c. Lia does not like chocolate.
 d. Lia loves her birthday.
 e. Lia doesn't enjoy birthdays.

11. Oscar Wilde said: "I always pass on good advice. It is the only thing to do with it. It is never of any use to oneself." Oscar Wilde was a great Irish writer best known for his sophisticated plays and witty remarks.

   Based on the paragraph, which of the following can we conclude is true?

 f. Oscar Wilde felt that a sophisticated play must always include good advice.
 g. Oscar Wilde believed that good advice must always be witty.
 h. Oscar Wilde didn't believe that taking good advice can be useful to oneself.
 j. Oscar Wilde listened to good advice whenever possible.
 k. Oscar Wilde wrote many witty plays.

12. Read the following statements:

    1. Gloria is younger than Francesca.

    2. Yvonne is older than Gloria.

    3. Yvonne is older than Francesca.

    If the first two statements are true, the third statement is

    a. true.

    b. false.

    c. probably false.

    d. partly true.

    e. impossible to determine from the information given.

13. Mike's dog, Rover, eats 10 pounds of dog food a week. Pete's dog, Fido, only eats 8 pounds of dog food a week. Rover is bigger than Fido.

    Which of the following is true according to the information given?

    f. Rover is overweight.

    g. Rover eats more than Fido.

    h. Big dogs eat more than small dogs.

    j. Mike and Pete are friends.

    k. Fido is older than Rover.

14. Mr. Brown, a popular local baker, offers a loaf of bread free of charge to every one of his customers who believe the bread they bought at Mr. Brown's bakery was stale. Mr. Brown's bread is Gabriela's favorite.

    Which of the following do we know is true?

    a. Gabriela is a customer of Mr. Brown.

    b. Gabriela is the only customer of Mr. Brown.

    c. Gabriela is Mr. Brown's favorite customer.

    d. Gabriela only buys bread at Mr. Brown's bakery.

    e. Gabriela always buys two loaves of bread at a time.

15. Dave earned $52 for three hours' work. Mark earned $60 for the work that he did. Donna earned $76 at her job.

    From the facts given, which of the following can we conclude is true?

    f. Donna makes more per hour than Mark.

    g. Mark earned more than Dave.

    h. Women earn more than men.

    j. Dave's work is inferior.

    k. It cannot be determined from the facts given.

Reading Comprehension

This section tests your reading comprehension—your ability to understand what you read. Read each passage carefully and answer the questions that follow it. If necessary, you can reread the passage to be certain of your answers. Remember that your answers must be based only on information that is actually in the passage.

Read the following passage and answer Questions 16 through 20.

Among the most popular museum exhibits in the world are those showcasing the treasures of ancient Egypt and Mesopotamia. In recent times, millions of people have viewed with wonder temple fragments, tomb contents, and the depiction of various deities. Yet few people today are aware that at the same time that Egypt and Mesopotamia were flourishing, a little-known city on the Indian subcontinent, called Harappa, was becoming the trading center of the known world.

    The area in which Harappa was located is today the Indus Valley of Pakistan and northwest India. Archeologists working in this area have unearthed a city that appears to have been orderly and functional, with little in the way of the opulent temples or lavish tombs that were a hallmark of Egypt and Mesopotamia. Instead, Harappa and other cities in the region were built for commerce. Although it probably began as an agricultural village around 3300 BCE, excellent farming land provided a reliable food source that allowed the village to thrive. Over the years, it transformed into an urban trading center because of its location near the crossroads of several major trade routes.

One indication of Harappa's importance as a trading center is the way its city wall was constructed. Because the single wall around Harappa had no moat, it was probably not built for defense. Archeologists believe that the people of Harappa probably never engaged in war; scientists working in the area have uncovered no evidence that the Harappans even had a standing army.

The main gate to the city, however, was nine feet wide, just wide enough for one ox cart to go through. Stone weights found at the gate indicated that taxes were being levied on goods that entered the city. Inside the city, archeologists uncovered elaborately carved stone seals that were used to mark exported goods. Unfortunately, no one has been able to decipher the inscriptions on those seals. Until the writing can be interpreted, much about the city, including the identities of the aristocracy of Harappa, will remain unknown.

Although the culture centered around Harappa thrived for many hundreds of years, between 1900 and 1700 BCE it showed signs of collapse. Farms were no longer productive, and trading networks broke down. One theory is that the Indus River shifted, flooding many of the agricultural villages around Harappa. This flooding may have affected the location of trade routes as well. This disruption of Harappa's

support system and main source of revenue probably led to the eventual disappearance of the entire culture.

16. Which of the following best tells what this passage is about?
    a. a once-thriving trade center on the Indian subcontinent
    b. the culture of Harappa compared to that of Egypt
    c. how the city of Harappa collapsed between 1900 and 1700 BCE
    d. how Harappa, Egypt, and Mesopotamia were once connected
    e. to show that lavish temples and tombs were once associated with urban centers

17. Which of the following facts from the passage best supports the idea that Harappans did not engage in warfare?
    f. Harappa was both orderly and functional.
    g. Harappa had no lavish tombs or opulent temples.
    h. Harappa was surrounded by just a single wall, and it had no moat.
    j. Harappa's main gate was just nine feet wide.
    k. Harappan writing has yet to be interpreted.

18. The passage mentions that all the following were uncovered by archaeologists working in Harappa EXCEPT
    a. stone seals used to mark the city's exports.
    b. evidence of agricultural production.
    c. tombs where the Harappan aristocracy was buried.
    d. a gated wall that encircled the entire city.
    e. stone weights used to levy the area's taxes.

19. What does the passage suggest as the most likely reason that Harappa went from an agricultural community to an urban center?
  f. After 3300 BCE, its farms were no longer productive.
  g. The Indus River shifted, flooding many agricultural villages.
  h. Agricultural output demanded commercial development.
  j. The local aristocracy decided to develop an export business.
  k. It was near the crossroads of several major trade routes.

20. Which of the following statements is supported by information given in the passage?
  a. The residents of Harappa paid income taxes.
  b. Harappa both imported and exported goods.
  c. Harappa's leaders were generous and kind.
  d. Harappa lacked an irrigation system.
  e. Farming in Harappa was more advanced than farming in Egypt.

Read the following passage and answer Questions 21 through 27.

Today's Olympic games are quite different from the original Olympics of nearly 3,000 years ago. The first recorded Olympic Games were held in 776 BC, and consisted of one event: a great footrace of about 200 yards. The race was held just outside the little town of Olympia in Greece.

From the date of that first game, the Greeks began to keep their calendar by Olympiads, the four-year spans between the celebrations of the famous games. The first Olympiad lasted from the summer of 776 BC to the summer of 772 BC. Today the term Olympiad refers to the period beginning in January of the years that the Summer Olympics will be held. An Olympiad still lasts for four years.

Greek women were not only forbidden to participate in the Olympic games, they were forbidden to even watch them. So the women held games of their own. The women's games, called the Heraea, were dedicated to the goddess Hera. The first Heraea consisted of footraces, like the men's games. The women's games continued to be held every four years, but they had fewer events than the Olympics.

Winning was of prime importance in both the male Olympiads and the female Heraea festival. The winners of the Olympics and of the Heraea were crowned with chaplets of wild olive, and in their home city-states, male champions were also awarded numerous honors, valuable gifts, and privileges.

The modern Olympic games, which started in Athens in 1896, are the result of the devotion of French educator Baron Pierre de Coubertin. He believed that because young people and athletics have gone together through the ages, putting education and athletics together might be a way to work toward better international understanding. Since then, the games have been held in cities throughout the world, with the goal of fostering world cooperation and athletic excellence.

What started in a small town in Greece has become one of the most well-known and honored events in the world. Over the years, many kinds of events have been added to the footraces started by the Greeks.

21. Which of the following best tells what this passage is about?
  a. Greece is where the Olympics began.
  b. how athletes trained for the Olympics
  c. the history of the Olympics
  d. The Olympics encourage world cooperation and athletic excellence.
  e. The Olympics started in Athens in 1896.

22. Today's Olympics are similar to the games of 3,000 years ago because
  f. women participate.
  g. there is more than one event.
  h. more than one country competes.
  j. the games are fiercely competitive.
  k. they have footraces.

23. How many years is an Olympiad?
  a. two
  b. four
  c. six
  d. one
  e. three

24. Which is NOT mentioned as a goal of today's Olympics?
  f. world cooperation
  g. putting education and athletics together
  h. international understanding
  j. winning awards
  k. athletic excellence

25. In the original Olympics, Greek women were not allowed to participate or watch. They responded by
  a. watching from afar.
  b. participating in disguise.
  c. conducting their own games.
  d. staying home.
  e. rejecting the idea of competitive games.

26. In the early days of the Olympics, male champions
  f. were given their freedom.
  g. were the strongest in the country.
  h. stayed away from the female champions.
  j. were awarded valuable gifts.
  k. were barred from future competitions.

27. The third paragraph describes
  a. women's response to the Olympics.
  b. the goals of ancient Greece's Olympics.
  c. what the winners of the Olympics receive.
  d. why athletic excellence is important.
  e. how the Olympics got started.

Read the following passage, then answer Questions 28 through 33.

Produced in 1959, Lorraine Hansberry's play, *A Raisin in the Sun*, was a quietly revolutionary work that depicted African-American life in a new and realistic way. The play made her the youngest American, the first African American, and the fifth woman to win the New York Drama Critic's Circle Award for Best Play of the Year. In 1961, it was produced as a film starring Sydney Poitier and has since become a classic, providing encouragement for an entire generation of African-American writers.

Hansberry was not only an artist but also a political activist and the daughter of activists. Born in Chicago in 1930, she was a member of a prominent family devoted to civil rights. Her father was a successful real estate broker, who won an anti-segregation case before the Illinois Supreme Court in the mid-1930s, and her uncle was a Harvard professor. In her home, Hansberry was privileged to meet many influential cultural and intellectual leaders.

The success of *A Raisin in the Sun* helped gain an audience for her passionate views on social justice. It mirrors one of Hansberry's central artistic efforts, that of freeing many people from the smothering effects of stereotyping by depicting the wide array of personality types and aspirations that exist within one Southside Chicago family. *A Raisin in the Sun* was followed by another play, produced in 1964, *The Sign in Sidney Brustein's Window*. This play is about an intellectual in

Greenwich Village, New York City, a man who is open-minded and generous of spirit who, as Hansberry wrote, "cares about it all. It takes too much energy not to care."

Lorraine Hansberry died on the final day of the play's run on Broadway. Her early death, at the age of 34, was unfortunate, as it cut short a brilliant and promising career, one that, even in its short span, changed the face of American theater. After her death, however, her influence continued to be felt. A dramatic adaptation of her autobiography, *To Be Young, Gifted, and Black*, consisted of vignettes based on Hansberry's plays, poems, and other writings. It was produced Off-Broadway in 1969 and appeared in book form the following year. Her play, *Les Blancs*, a drama set in Africa, was produced in 1970; and *A Raisin in the Sun* was adapted as a musical, *Raisin*, and won a Tony award in 1973.

Even after her death, her dramatic works have helped create an audience for her essays and speeches on topics ranging from world peace to the evils of the mistreatment of minorities. Hansberry was a woman, much like the characters in her best-known play, who was determined to be free of racial, cultural, or gender-based constraints.

28. Which of the following best tells what this passage is about?
a. Hansberry wrote excellent plays that had both artistic and political influence.
b. Hansberry's parents deserve credit for raising a successful daughter.
c. These are Hansberry's best-known works.
d. If one is raised in a well-educated family, one is likely to succeed.
e. Hansberry had a difficult struggle, but ultimately succeeded as a young female writer.

29. The writer of this passage suggests that Hansberry's political beliefs originated with her experience as
f. the daughter of politically active parents.
g. a successful playwright in New York.
h. a resident of Southside Chicago.
j. an intellectual in Greenwich Village.
k. a successful real estate broker.

30. Paragraph 3 states that Hansberry's main purpose in writing *A Raisin in the Sun* was to
a. win her father's approval.
b. break down stereotypes.
c. show people how interesting her own family was.
d. earn the right to produce her own plays.
e. influence members of the Illinois Supreme Court.

31. Which of the following best explains the author's reason for including paragraphs 4 and 5?
f. The cities of Chicago and New York have similar problems.
g. The civil rights struggle continued even after Hansberry died.
h. Hansberry actually wrote more poems and essays than she did plays.
j. *A Raisin in the Sun* was more successful after Hansberry's death than it was before she died.
k. Hansberry's work continued to influence people even after her death.

32. According to the passage, how many women had won the New York Drama Critic's Circle Award for Best Play of the Year before Lorraine Hansberry did?
a. none
b. one
c. two
d. four
e. five

33. According to the passage, which of the following dramatic works was based most directly on Hansberry's life?

  f. *A Raisin in the Sun*

  g. *Les Blancs*

  h. *The Sign in Sidney Brustein's Window*

  j. *To Be Young, Gifted, and Black*

  k. *Southside Chicago*

Read the following passage, then answer questions 34 through 38.

It looked like a good thing: but wait till I tell you. We were down South, in Alabama—Bill Driscoll and myself—when this kidnapping idea struck us. . . .

 We selected for our victim the only child of a prominent citizen named Ebenezer Dorset. The father was respectable and tight. . . . The kid was a boy of ten. Bill and me figured that Ebenezer would melt down for a ransom of two thousand dollars to a cent. . . .

 One evening after sundown, we drove in a buggy past old Dorset's house. The kid was in the street, throwing rocks at a kitten on the opposite fence.

 "Hey, little boy!" says Bill, "would you like to have a bag of candy and a nice ride?"

 The boy catches Bill neatly in the eye with a piece of brick.

 "That will cost the old man an extra five hundred dollars," says Bill.

 The boy put up a fight, but, at last, we got him down in the bottom of the buggy and drove away. We took him up to the cave. . . .

 Yes, sir, that boy seemed to be having the time of his life. He immediately christened me Snake-eye, the Spy, and announced that I was to be broiled at the stake at the rising of the sun.

[The boy calls himself "Red Chief, the terror of the plains."]

 "Red Chief," says I to the kid, "would you like to go home?"

 "Aw, what for?" says he. "I don't have any fun at home. I hate to go to school. . . . I never had such fun in all my life."

 We went to bed around eleven o'clock. I fell into a troubled sleep.

 Just at daybreak, I was awakened by a series of awful screams from Bill. Red Chief was sitting on Bill's chest, with one hand twined in Bill's hair. In the other he had the sharp case-knife we used for slicing bacon; and he was trying to take Bill's scalp. . . .

 I got the knife away from the kid and made him lie down again. . . . I dozed off for a while, but along toward sun-up I remembered that Red Chief had said I was to be burned at the stake at the rising of the sun. I wasn't nervous or afraid; but I sat up and lit my pipe and leaned against a rock.

 "What you getting up so soon for, Sam?" asked Bill.

 "Me?" says I. "Oh, I got a kind of a pain in my shoulder. I thought sitting up would rest it."

 "You're a liar!" says Bill. "You're afraid. You was to be burned at sunrise, and you was afraid he'd do it. And he would, too, if he could find a match."

 "Ain't it awful, Sam?" says Bill. "Do you think anybody will pay out money to get a little imp like that back home?"

—O. Henry, from "The Ransom of Red Chief" (1907)

34. Which of the following best tells what this passage is about?
  a. Two kidnappers gain a lot of money from their misdeeds.
  b. A brave little boy escapes from two dangerous kidnappers.
  c. Two kidnappers end up fighting with each other.
  d. Two kidnappers become fearful of their victim.
  e. Two would-be kidnappers lose their victim before they can collect the ransom.

35. Why do Sam and Bill want to kidnap the son of Ebenezer Dorset?
  f. They know that Ebenezer will pay them a ransom just to avoid publicity.
  g. They have failed to kidnap any other children from Summit.
  h. They need $100 and know that Ebenezer will pay that sum.
  j. Ebenezer's son is sickly and weak and will be easy to kidnap.
  k. They think that he is wealthy enough to pay a good deal of money to get his only son back.

36. Why doesn't the kidnapping victim want to go home?
  a. He is afraid his father will punish him.
  b. He wants the sweets that the kidnappers offered him.
  c. He believes he can escape whenever he wants to.
  d. He wants to keep the ransom for himself.
  e. He finds being kidnapped more fun than going to school.

37. How is Red Chief different from a typical kidnapping victim?
  f. Red Chief acts scared and pleads to go home.
  g. Red Chief scares and hurts his kidnappers.
  h. Red Chief doesn't like being kept hidden in a cave.
  j. Red Chief is afraid his father will not be able to meet the ransom.
  k. Red Chief is afraid he will be killed by his kidnappers.

38. Later in the story, after another day spent with Red Chief, Sam and Bill write a ransom note to Ebenezer. They sign the note, "Two Desperate Men." Normally criminals use this phrase to indicate that they will do anything—even kill—to get what they want. Based on the passage, what do Sam and Bill most likely mean by that signature?
  a. They cannot deal with Red Chief and are desperate to get out of the situation.
  b. They will kill Red Chief if they don't get what they want.
  c. They are wanted for crimes in other states.
  d. They are hungry and urgently need food.
  e. A storm is coming and they need more shelter than the cave can offer.

Read the following passage, then answer Questions 39 through 45.

Millions of people in the United States are affected by eating disorders. More than 90% of those afflicted are adolescents or young adult women. Although all eating disorders share some common manifestations, anorexia nervosa, bulimia nervosa, and binge eating disorder each have distinctive symptoms and risks.

People who intentionally starve them-selves (even while experiencing severe hunger pains) suffer from anorexia nervosa. The disorder, which usually begins around the time of puberty, involves extreme weight loss to at least 15% below the individual's normal body weight. Many people with the disorder look emaciated but are convinced they are overweight. In patients with anorexia nervosa, starvation can damage vital organs such as the heart and brain. To protect itself, the body shifts into slow gear: menstrual periods stop, blood pressure rates drop, and thyroid function slows. Mild anemia, swollen joints, reduced muscle mass, and light-headedness also commonly occur in anorexia nervosa.

Anorexia nervosa sufferers can exhibit sudden angry outbursts or become socially withdrawn. One in ten cases of anorexia nervosa leads to death from starvation, cardiac arrest, other medical complications, or suicide.

People with bulimia nervosa consume large amounts of food and then rid their bodies of the excess calories by vomiting, abusing laxatives or diuretics, taking enemas, or exercising obsessively. Some use a combination of all these forms of purging. Individuals with bulimia who use drugs to stimulate vomiting, bowel movements, or urination may be in considerable danger, as this practice increases the risk of heart failure.

Because many individuals with bulimia binge and purge in secret and maintain normal or above normal body weight, they can often successfully hide their problem for years. But bulimia nervosa patients—even those of normal weight—can severely damage their bodies by frequent binge eating and purging. In rare instances, binging causes the stomach to rupture; purging may result in heart failure due to loss of vital minerals such as potassium. As

in anorexia nervosa, bulimia may lead to irregular menstrual periods. As with anorexia nervosa, bulimia typically begins during adolescence. Eventually, half of those with anorexia nervosa will develop bulimia.

Binge eating disorder is found in about 2% of the general population. This disorder differs from bulimia because its sufferers do not purge. Individuals with binge eating disorder eat large quantities of food and do not stop until they are uncomfortably full. Most sufferers are overweight or obese and have a history of weight fluctuations. As a result, they are prone to the serious medical problems associated with obesity, such as high choles-terol, high blood pressure, and diabetes. Obese individuals also have a higher risk for gallbladder disease, heart disease, and some types of cancer.

39. Which of the following best tells what this passage is about?
   a. the connections between eating disorders and the loss of control
   b. Eating disorders can become a major problem for adolescents or young adult women.
   c. Eating disorders affect body weight in various ways.
   d. There are some similarities between bulimia and anorexia nervosa.
   e. an overview of eating disorders and their consequences

40. Fatalities occur in what percent of people with anorexia nervosa?
   f. 2%
   g. 10%
   h. 15%
   j. 30%
   k. 45%

41. Which of the following consequences do all the eating disorders mentioned in the passage have in common?
   a. heart ailments
   b. stomach rupture
   c. swollen joints
   d. diabetes
   e. asthma

42. According to the passage, people with binge eating disorders are prone to all the following EXCEPT
   f. obesity.
   g. gallbladder disease.
   h. low blood pressure.
   j. high cholesterol.
   k. diabetes.

43. Which of the following is NOT a statement from the passage about people with eating disorders?
   a. People with anorexia nervosa commonly have a blood-related deficiency.
   b. People with anorexia nervosa perceive themselves as overweight.
   c. The female population is the primary group affected by eating disorders.
   d. Binging and purging in secret is typical of those with bulimia.
   e. Fifty percent of people with bulimia also develop anorexia nervosa.

44. People who have an eating disorder but nevertheless appear to be of normal weight are most likely to have
   f. obsessive-compulsive disorder.
   g. bulimia nervosa.
   h. binge eating disorder.
   j. anorexia nervosa.
   k. binge eating disorder and anorexia nervosa.

45. According to the passage, which of the following is true of bulimia patients?
   a. They may demonstrate unpredictable social behavior.
   b. They often engage in compulsive exercise.
   c. They are less susceptible to dehydration than are anorexia patients.
   d. They frequently experience stomach ruptures.
   e. They are usually very outgoing.

Part 2—Math

The Math Test includes 50 multiple choice questions covering content in the following areas:

- basic math
- percentages, fractions, decimals, averages
- pre-algebra
- algebra
- substitution
- factoring
- coordinate graphing
- geometric principles
- logic
- word problems

Solve each problem and select the best answer from the choices given. It is important to keep in mind that:

- Formulas and definitions of mathematical terms and symbols are not provided.
- Diagrams other than graphs are not necessarily drawn to scale. Do not assume any relationship in a diagram unless it is specifically stated or can be figured out from the information given.
- You can assume that a diagram is in one plane unless the problem specifically states that it is not.

- Graphs are drawn to scale. Unless stated otherwise, you can assume relationships according to appearance. For example, (on a graph) lines that appear to be parallel can be assumed to be parallel; likewise for concurrent lines, straight lines, collinear points, right angles, and so on.
- You must reduce all fractions to lowest terms.

46. Examine (A), (B), and (C) and find the best answer.

 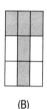 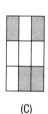

    (A)   (B)   (C)

  a. (A) is more shaded than (B).
  b. (B) is less shaded than (C).
  c. (A) and (B) are equally shaded.
  d. (B) and (C) are equally shaded.
  e. (A) and (C) are equally shaded.

47. Which of the following decimals has the *least* value?
  f. 0.00120
  g. 0.01020
  h. 0.012
  j. 0.12
  k. 0.00102

48. The perimeter of a triangle is 25 inches. If side a is twice side b, which is $\frac{1}{2}$ side c, what is the length of side b?
  a. 5
  b. 8
  c. 10
  d. 12
  e. 15

49. At a slumber party, the following names are put into one hat: Anna, Beth, Carla, and Diana. A second hat contains the names Evelyn, Francine, Gillian, Hallie, and Ida. If one winner will be selected from each hat, what is the probability that Beth and Gillian will both win?
  f. $\frac{1}{9}$
  g. $\frac{1}{2}$
  h. $\frac{1}{16}$
  j. $\frac{1}{20}$
  k. $\frac{1}{24}$

50. Ron is half as old as Sam, who is three times as old as Ted. The sum of their ages is 55. How old is Ron?
  a. 5
  b. 8
  c. 10
  d. 12
  e. 15

51. Which of the following groups is NOT a subset of the rational numbers?
  f. irrational numbers
  g. natural numbers
  h. integers
  j. whole numbers
  k. negative numbers

52. The sun is approximately 9.3×10^7 miles from the earth. Which expression below is equivalent to 9.3×10^7 miles?
  a. 93,000 miles
  b. 930,000 miles
  c. 93 million miles
  d. 930,000,000 miles
  e. 9.3 million miles

53. Here are test scores for five students: 79, 79, 81, 82, and 84. Which statement below is true?
  f. The mean and median scores are the same.
  g. The mode score is greater than the median score.
  h. The mean score is greater than the median score.
  j. The mean score is less than the median score.
  k. The mode score is the same as the median score.

54. The drawing below represents a deck that the Joneses are building. In the drawing, one-half inch represents three feet. What are the actual dimensions of the deck?

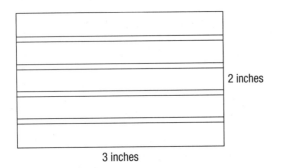

2 inches

3 inches

  a. 3 feet by 4.5 feet
  b. 6 feet by 9 feet
  c. 8 feet by 6 feet
  d. 7 feet by 10.5 feet
  e. 12 feet by 18 feet

55. What number is 7 less than $\frac{3}{4}$ of 20?
  f. −2
  g. −8
  h. 8
  j. 10
  k. 13

56. All the rooms on the top floor of a government building are rectangular, with 8-foot ceilings. One room is 9 feet wide by 11 feet long. What is the combined area of the four walls, including doors and windows?
  a. 99 square feet
  b. 160 square feet
  c. 320 square feet
  d. 72 square feet
  e. 792 square feet

57. A ticket to an evening movie at the Bijou costs $7.50. The cost of popcorn at the concession stand is 80% of the cost of a ticket. How much does popcorn cost?
  f. $6.00
  g. $6.50
  h. $6.70
  j. $7.00
  k. $5.50

Use the following graph to answer Questions 58 through 60.

The graph shows the yearly electricity usage for Finnigan Engineering Inc. over the course of four years for three departments.

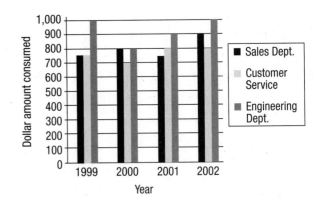

58. How much greater was the electricity cost for sales during the year 1999 than the electricity cost for customer service in 2000?

- **a.** $250
- **b.** $200
- **c.** $150
- **d.** $100
- **e.** $50

59. Which of the following statements is supported by the data?

- **f.** The sales department showed a steady increase in the dollar amount of electricity used during the 4-year period.
- **g.** On average, the customer service department used more electricity than the sales department.
- **h.** The customer service department showed a steady increase in the dollar amount of electricity used during the 4-year period.
- **j.** The engineering department showed a steady increase in the dollar amount of electricity used from 2000 to 2002.
- **k.** None of the above are supported by the data.

60. What was the percent decrease in electricity usage (in dollar amount) from 1999 to 2000 for the engineering department?

- **a.** 25%
- **b.** 20%
- **c.** 15%
- **d.** 10%
- **e.** 5%

61. The coordinates of the vertices of ABC are $A(1,0)$; $B(6,0)$; and $C(6,12)$. What is the length of $\overline{AC}$?

- **f.** 17
- **g.** 14.2
- **h.** 8.5
- **j.** 13
- **k.** 15

62. Billy's bank is the shape of a cube. If the length of one edge of the bank is 5 inches, what is the volume of the bank? (The volume of a cube is equal to the length of one edge of the cube to the third power.)

- **a.** 15 in.^{3}
- **b.** 25 in.^{3}
- **c.** 125 in.^{3}
- **d.** 325 in.^{3}
- **e.** 500 in.^{3}

63. Solve for x in the following equation:
$$1.5x - 7 = 12.5.$$

- **f.** 29.25
- **g.** 23
- **h.** 19.5
- **j.** 13
- **k.** 5.5

64. $-12\frac{2}{7} - (-3\frac{4}{7}) =$

- **a.** $-15\frac{6}{7}$
- **b.** $-15\frac{2}{7}$
- **c.** $-9\frac{2}{7}$
- **d.** $-8\frac{5}{7}$
- **e.** $-6\frac{5}{7}$

65. Look at this series: $1, \frac{7}{8}, \frac{3}{4}, \frac{5}{8}, \ldots$ What number should come next?

- **f.** $\frac{2}{3}$
- **g.** $\frac{9}{4}$
- **h.** $\frac{1}{2}$
- **j.** $\frac{3}{8}$
- **k.** $\frac{1}{4}$

66. Six friends agree to evenly split the cost of gasoline on a trip. Each friend paid $37.27. What was the total cost of gas?

- **a.** $370.27
- **b.** $223.62
- **c.** $286.56
- **d.** $314.78
- **e.** $262.78

67. Use the following to answer Question 67.

Zone	Weight	Shipping Cost
1	fewer than 50 lbs. more than 50 lbs.	$1.50 $2.50
2	fewer than 50 lbs. more than 50 lbs.	$3.00 $4.50
3	fewer than 50 lbs. more than 50 lbs.	$4.00 $5.50

How much more will Sherry pay in shipping costs to ship packages B and D compared to shipping packages A and C?

Package A: 44 lbs. Shipped to Zone 2.
Package B: 83 lbs. Shipped to Zone 3.
Package C: 12 lbs. Shipped to Zone 1.
Package D: 75 lbs. Shipped to Zone 3.

f. $6.50
g. $5.50
h. $4.50
j. $3.50
k. $3.00

68. One evening, Super Tunes 96.6 FM sponsored a phone pledge-a-thon. The average caller pledged a donation of $796. If there were 182 callers, how much money was raised in the pledge-a-thon?
a. $8,756
b. $15,920
c. $36,256
d. $144,872
e. $217,882

69. This map shows the northern part of Italy. The scale on the map is 1 unit:100 km. Which is the approximate distance, in kilometers, between Pisa and Venice?

f. 100 km
g. 250 km
h. 500 km
j. 800 km
k. 1000 km

70. Simplify the following algebraic expression:

$$5x^2 + 9x - 3 - 3(x^2 - x + 4)$$

a. $2x^2 + 12x - 15$
b. $2x^2 + 6x + 9$
c. $2x^2 + 12x - 10$
d. $2x^2 + 5x - 15$
e. $2x^2 - 5x - 10$

71. Which is the best estimate of $\sqrt{181}$?
f. 9
g. 11
h. 12
j. 13
k. 14

72. Find the area of the shape.

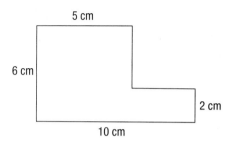

    a. 60 cm^2
    b. 23 cm^2
    c. 50 cm^2
    d. 58 cm^2
    e. 40 cm^2

73. The heights of 8 students were measured in feet and then ordered from least to greatest on a number line. Their heights were as follows: $5\frac{1}{2}$ ft, 5.25 ft, $5\frac{1}{6}$ ft, 5.8 ft, $5\frac{3}{4}$ ft, 5.4 ft, 5.2 ft, and $5\frac{7}{8}$ ft. Which number was third on the number line?

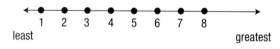

    f. $5\frac{1}{2}$
    g. 5.25
    h. $5\frac{3}{4}$
    j. 5.4
    k. 5.8

74. On Sunday afternoon, a crowd of 57,800 people attended a football game. At halftime, the home team was trailing by a score of 20 to 0, and 5,167 people left the stadium to go home. During the second half, more people left the stadium before the game ended. The stadium has 40 sections. Each section has 50 rows, and every row has 30 seats.

    Which problem CANNOT be solved using the information given?

    a. the total number of seats in the stadium
    b. the number of seats in each section of the stadium
    c. the number of people who are in the stadium before the football game ends
    d. the number of people who are in the stadium as the second half of the football game begins
    e. answers **a** and **b**

75. Anne has two containers for water: a rectangular plastic box with a base area of 16 square inches, and a cylindrical container with a radius of 2 inches and a height of 11 inches. If the rectangular box is filled with water 9 inches from the bottom, and Anne pours the water into the cylinder without spilling, which of the following will be true? Use the following formulas for volume: (Volume of a cylinder) = $(\pi r^2)(h)$; (Volume of a rectangular prism) = (Area of Base)(h).

    f. The cylinder will overflow.
    g. The cylinder will be exactly full.
    h. The cylinder will be filled to an approximate level of 10 inches.
    j. The cylinder will be filled to an approximate level of 12 inches.
    k. The cylinder will be filled to an approximate level of 8 inches.

76. Simplify the following expression: $7^{-3} \times 7^2 \times 7^4 \times 7^{-4}$

   a. $\frac{1}{7}$

   b. 49

   c. 7

   d. 1

   e. 7^{96}

77.

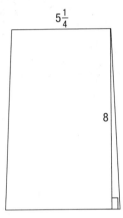

The formula for the area of a trapezoid is $A = \frac{1}{2}h(b_1 + b_2)$. If the area of the trapezoid is 45, which expression below gives the length of the other base?

   f. $\dfrac{2(45)}{5\frac{1}{4}} - 8$

   g. $\dfrac{2(45)}{8} - 5\frac{1}{4}$

   h. $\dfrac{2(45)}{5\frac{1}{4}} - \dfrac{2(45)}{8}$

   j. $\dfrac{(5\frac{1}{4})(45)}{8} - \dfrac{2(45)}{8}$

   k. $\dfrac{(5\frac{1}{4})(45)}{8}$

78. Which of the following is equivalent to $2x(3xy + y)$?

   a. $6(x^2)y + 2xy$

   b. $6xy + 2xy$

   c. $5x2y + 2x + y$

   d. $6xy + 2x + y$

   e. $3xy + 2x + y$

79. Marvelous Maids pledged $5.25 for each house they cleaned during the third week of May. Their total donation was $1,291.50. How many houses did they clean during the third week of May?

   f. 25

   g. 26

   h. 245

   j. 246

   k. 6,780

80. Point P shown on the following graph has coordinates of $(1,-3)$. A second point, Q, is located two units to the left of point P and five units above it. Which ordered pair represents the coordinates of point Q?

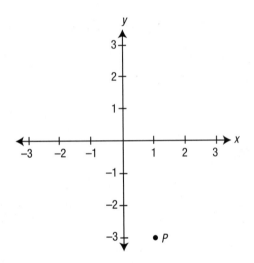

   a. $(-2,5)$

   b. $(2,-1)$

   c. $(-1,2)$

   d. $(2,1)$

   e. $(-2,-1)$

81. Samantha earned $576 over a two-week period. She worked 32 hours each of the two weeks. She earns the same amount of money for each hour that she works. What is her hourly rate of pay?
   f. $7
   g. $8
   h. $9
   j. $12
   k. $18

82. Which is the *best* unit of measurement to measure the length of a book?
   a. centimeters
   b. kilometers
   c. yards
   d. meters
   e. miles

83. In Derrick's sock drawer, there were 5 blue socks, 12 white socks, and 7 black socks. He reached into the drawer without looking and pulled out the first sock he touched. What were the odds he pulled out a blue sock?
   f. 5:19
   g. 19:5
   h. 5:24
   j. 24:5
   k. 19:24

84. Michael takes three minutes to read two pages in his book. At this rate, how long will it take him to read a 268-page book?
   a. 6 hours and 42 minutes
   b. 13 hours and 24 minutes
   c. 9 hours and 12 minutes
   d. 8 hours and 56 minutes
   e. 5 hours and 2 minutes

85. The bar graph shows the number of movie tickets purchased each day for a week.

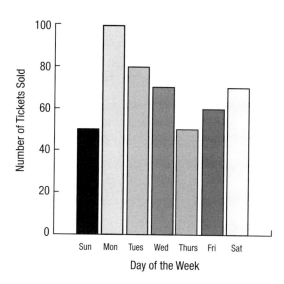

Movie Ticket Sales for the Week

How many total tickets were sold during the entire week?

   f. 350
   g. 380
   h. 410
   j. 430
   k. 480

86. A package contains twelve chocolate bars. A group of friends buys one package and shares the chocolate bars evenly among themselves without breaking up any of the bars. Which of the following could NOT be the number of friends in the group?
   a. 2
   b. 4
   c. 6
   d. 8
   e. 12

87. A drawing you made measures 8 inches wide by 12 inches high. You want to include the drawing in a report you are preparing, but the drawing can only be 5 inches wide. You decide to make a reduction of the original drawing. What is the height of the reduced drawing if its width is 5 inches?

   f. $7\frac{1}{2}$ inches

   g. $3\frac{1}{3}$ inches

   h. $19\frac{1}{5}$ inches

   j. 9 inches

   k. $8\frac{3}{4}$ inches

88. Choose the statement that must be true if it is given that:

- If Jim practices his musical instrument, then he makes it to Carnegie Hall and becomes a star.
- Jim practices his musical instrument.

   a. Jim is not a star.

   b. Jim makes it to Carnegie Hall and becomes a star.

   c. Jim doesn't make it to Carnegie Hall.

   d. Jim wastes his time playing his instrument.

   e. Jim doesn't practice his instrument enough.

89. Two numbers have a sum of 98. One number is 6 more than 3 times the other number. What is the smaller number?

   f. 23

   g. 52

   h. 75

   j. 98

   k. 300

90. The perimeter of a square is 24 inches. What is its area?

   a. 144 in²

   b. 576 in²

   c. 24 in²

   d. 16 in²

   e. 36 in²

91. Oliver made the following figure by cutting and folding a piece of cardboard. Which drawing shows the piece of cardboard Oliver used to make the figure?

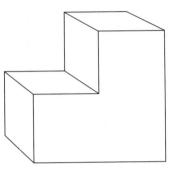

f.

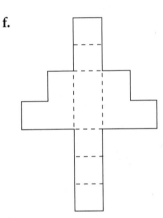

g.

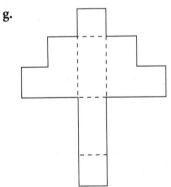

h.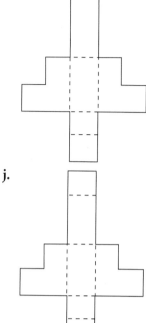

j.

k. None of the above

92. Jackie did a survey for her social studies project. She asked people if they thought that recycling is important. Of the people Jackie talked to, $\frac{7}{8}$ said that recycling is important. Only $\frac{1}{3}$ of the people who said recycling is important said that they also buy things made from recycled materials. What fraction of the people she talked to think recycling is important *and* buy things made from recycled materials?

a. $\frac{6}{5}$

b. $\frac{13}{14}$

c. $\frac{2}{6}$

d. $\frac{7}{24}$

e. $\frac{1}{24}$

93. In County A, 325 of every 1,000 people live in large cities. Which proportion can be used to determine x, the total number of people in County A who live in large cities, if the county's population is 970,000?

f. $\frac{x}{1,000} = \frac{325}{970,000}$

g. $\frac{325}{1,000} = \frac{x}{970,000}$

h. $\frac{x}{1,000} = \frac{1,000 - 325}{970,000}$

j. $\frac{1,000 - 325}{1,000} = \frac{x}{970,000}$

k. $\frac{x}{1,000} = \frac{970,000}{325}$

94. Which of the following expressions is the prime factorization of 60?

a. 2×30

b. $2 \times 2 \times 4 \times 5$

c. $5 \times 2 \times 3$

d. 10×6

e. $2 \times 2 \times 3 \times 5$

95. What would the area of the triangle be if all its dimensions were one-third of those shown below?

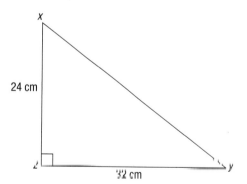

f. 3,456 cm^2

g. 256 cm^2

h. 384 cm^2

j. 96 cm^2

k. 42.67 cm^2

Answers

Paragraph 1 (R, T, U, S, Q)

The paragraph mentions "myths about bats" and sentence **R**, with an example of a myth, is the only one of these choices that could logically come next. Note that the word "however" indicates that sentence **U** must follow and contradict some other statement, so the best order is **T** and then **U**. Sentence **S** tells what bats *are* related to, and sentence **Q** follows with one similarity between bats and humans.

Paragraph 2 (S, R, T, Q, U)

The given sentence states that some authors live dangerous lives and sentences **R** and **S** give examples of two who lived dangerous lives, so **R** and **S** logically follow the introduction. The words "It is also true" make it clear that **R** should come after **S**. Sentences **T**, **Q**, and **U** all refer to a different type of writer. **T** states the contradiction, **Q** gives an example, and **U** gives more information about that example.

Paragraph 3 (S, U, T, Q, R)

Sentence **S** states the first thing to do. Sentence **U** (how to remove the old pot) logically comes before **T** (what to do once you have removed the old pot). Sentences **Q** and **R** are about putting the plant into the new pot; choosing a pot, **Q**, comes before lowering the plant into the pot, **R**.

Paragraph 4 (U, S, Q, T, R)

The given sentence announces that new carpeting will be installed and sentence **U** provides the guidelines for that installation. Sentences **Q** and **S** tell which areas will be carpeted. The world *also* in **Q** signals that it comes after **S**. Sentence **T** refers to both areas that will be carpeted. Sentence **R** gives an exception and mentions a third office area. Sentence **R** could not logically be placed anywhere before **T**, so it must come at the end of the paragraph.

Paragraph 5 (T, Q, U, R, S)

Sentence **T** states that the Ch'in Dynasty *first* unified the country. Sentence **Q** makes it clear what came second ("then . . . named himself"). Sentence **U** refers to "the new emperor" so it must come after **Q**. Sentence **R** states what the emperor did next. That leaves sentence **S**, telling what the emperor *also* did, to come at the end of the paragraph.

6. **a.** Since Maui is an island and islands are surrounded by water, Maui must be surrounded by water. There is not enough information to determine whether statements II and III are true.

7. **h.** To answer this, you need to find out what each letter represents. Note rule 3: the position of the letter in the code line does not tell you which word it represents. The words "Where is the" appear in all three sentences; the letters S, B, and G appear in all three sentences. The word "bus" appears in the last two sentences, and the letter **D** appears in the last two sentences. Therefore, **D** represents "bus."

8. **a.** S, B, and G appear in all three sentences, as do the words "where," "is," and "the." Since a letter is never in the same position as its word, "is" can never be the third letter in a sequence. So "is" cannot be letter **B**, leaving S and G. S appears as the first letter and the third letter, so S cannot be "where" or "the." So, S must be "is."

9. **f.** If George is sitting at Henry's left, George's seat is 252. The next seat to the left, then, is 251.

10. **b.** The first statement says that Lia celebrated her birthday; and the second statement says that she received many presents, including a box of chocolates. From this, we can conclude that Lia received a box of chocolates for her birthday (choice **b**). The other statements in the passage may or may not be true, but we cannot reach any conclusion about them based on the passage.

11. h. The only thing that can be concluded from Oscar Wilde's quote, for certain, is that he did not believe that taking good advice can be useful to oneself. Based only on the information given, we do not know what Oscar Wilde felt should be included in a play (choice **f**) or what constitutes good advice (choice **g**). Choices **j** and **k** are opinions, not facts stated in the paragraph.

12. e. Look at the information you know. Adding real world data can help you consider the problem:

Give Gloria an age, let's say 10.

Francesca is older than Gloria, so she's 11 or older.

Yvonne is older than Gloria, so she's also 11 or older.

This does not tell you whether Yvonne is older than Francesca.

So the statement "Yvonne is older than Francesca" might be true or false; it cannot be determined from the information given. The correct answer is choice **e**.

13. g. The only conclusion that we can draw from the information given is that Rover eats more than Fido. The other choices are not addressed.

14. a. The one fact we know for certain from the two statements is that Gabriela is a customer of Mr. Brown (choice **a**)—since his bread is her favorite. We know for a fact that Gabriela is not the only customer of Mr. Brown's (choice **b**) because we are told in the first statement that he is a popular local baker. Choices **c**, **d**, and **e** could be true, but they are not supported by the facts given.

15. g. Notice that we only know how many hours Dave worked—we are not told how long it took Mark and Donna to earn their money. The only thing that we can conclude for certain is that Mark made more than Dave.

16. a. This is the best choice because it correctly describes the main point of the passage. Although the passage does make a comparison between Harappa and Egypt (choice **b**), this is not the main purpose of the passage. Choices **c** and **d** are mentioned in the passage, but are not main points. Choice **e** is not supported by the information in the passage.

17. h. The third paragraph supports that point by stating that the single wall and lack of a moat around Harappa indicated that Harappa did not need a defense against invaders. The other choices do not support that inference.

18. c. Note that this question asks for the exception. The passage states that archaeologists did find stone seals (choice **a**), evidence of agricultural production (choice **b**), a gated wall (choice **d**), and stone weights used for taxes (choice **e**).

19. k. The closing sentence in the second paragraph supports this detail. Choices **f** and **g** reference information that details the overall collapse of the region rather than the transformation from farming center to trade center. Choice **h**, though logical, is unsupported by the information in the text. Choice **j** is also unsupported by the text.

20. b. Support for this answer is given in the fourth paragraph. Choice **a** is wrong because even though taxes were levied on goods entering Harappa, there is no evidence that those taxes were income taxes. There is no support in the passage for the other choices (**c**, **d**, or **e**).

21. c. This is the best choice because it refers to the entire passage. Choices **a**, **d**, and **e** are too limited. Choice **b** is not covered in the text.

22. k. The first paragraph mentions "a great footrace" in the ancient games, and footraces are mentioned again the final paragraph about the modern games. Choices **f, g,** and **h** are ways in which today's Olympic games differ from the old Olympic games. Choice **j** is not covered in the passage.

23. b. The second paragraph identifies an Olympiad as a "four-year span between the celebrations of the famous games."

24. j. Winning awards was mentioned as a goal of the Olympics in the past but not as a goal of the modern games. The other choices *are* mentioned as goals for the modern games.

25. c. The third paragraph states that women held games of their own, called the Heraea.

26. j. The fourth paragraph states that male champions were awarded numerous honors, valuable gifts, and privileges. There is nothing in the passage to suggest choices **a, b, c,** or **e.**

27. a. The best answer is "women's response to the Olympics." Choices **b, c,** and **e** are covered in other paragraphs. Choice **d** is not addressed directly in the passage.

28. a. The passage begins and ends with praise of Hansberry's works and influence. There is no support for choice **b.** Hansberry's works are summarized (choice **c**), but this is not the main purpose of the passage. Choice **d** is not necessarily true and is not in the passage. Lorraine Hansberry may have had a difficult struggle (choice **e**), but the struggle is not mentioned in the passage.

29. f. The first paragraph speaks of Hansberry being raised as the daughter of political activists. Choices **g** and **j** are related to her beliefs, but are not depicted as the origin of those beliefs. The passage does not say that Hansberry herself ever lived in Southside Chicago (choice **h**). Her father was a real estate broker; she was not (choice **k**).

30. b. This correct answer is clearly stated in the second sentence of paragraph 3. There is no support in the passage for the other choices.

31. k. Both paragraphs focus on how much Hansberry's work continued to be an influence even after she died. In paragraph 4 "Her influence continued to be felt," and in paragraph 5 "Even after her death, her dramatic works have helped create an audience." Choice **g** may be true, but this is not the main purpose of the paragraphs. There is no support for choices **f, h,** or **j.**

32. d. See the second sentence of the first paragraph. She was the fifth woman to win the award, which means there were four women before her.

33. j. See paragraph 4, which describes *To Be Young, Gifted, and Black* as a dramatic adaptation of an autobiography. Choice **f** is wrong because there is no support for the idea that *Raisin* is about Hansberry's family. *Les Blancs* is set in Africa, which rules out choice **g.** *The Sign in Sidney Brustein's Window* is about a man, which rules out choice **h.** Southside Chicago (choice **k**) is not the title of a dramatic work.

34. d. According to the passage, the kidnapping victim tries to scalp one kidnapper. This makes the other kidnapper nervous about the boy's threat to burn him at sunrise. Choice **c** might be true, but is not covered in the passage. None of the other answers is supported by the text.

35. k. The text clearly states that Ebenezer Dorset is a prominent citizen, which means that he is likely to be well off. The kidnappers believe he will pay a ransom of two thousand dollars, so choice **k** is the correct choice. None of the other answers are supported by the text.

36. e. Red Chief tells his kidnappers, "I hate to go to school. . . . I never had such fun in all my life." There is no support for any of the other answers in the passage.

37. g. Red Chief does not act like the usual kidnapping victim. Rather than being frightened and mistreated, he scares his captors and inflicts pain on Bill. So choice **g** is the correct answer. None of the other choices is supported by the text. They describe typical kidnapping victims, not Red Chief.

38. a. From the events in the story, you can infer that Sam and Bill are desperate to get out of the situation. The two never threaten to kill Red Chief (choice **b**). Choices **c, d,** and **e** are not supported by evidence in the text.

39. e. This is the only statement that is general enough to cover the entire passage. The other choices only mention details from the passage.

40. g. See the third paragraph: "One in ten" (10% of) cases of anorexia end in death. The other answers are not accurate, according to the passage.

41. a. See the second paragraph for reference to heart problems with anorexia, the fourth and fifth paragraphs for discussion of heart problems with bulimia, and the last paragraph, where heart disease is mentioned, as a risk in obese people who suffer from binge eating disorder.

42. h Near the end of the last paragraph, the passage indicates that binge eating disorder patients experience high blood pressure, not low blood pressure.

43. e. Note that the question asks for the statement NOT included. In choice **e**, the passage states that it is actually the other way around: 50% of people with anorexia develop bulimia, as stated near the end of the fifth paragraph. The other statements are included in the passage.

44. g. The first sentence of the fifth paragraph tells us that bulimia sufferers are often able to keep their problem a secret, partly because they maintain a normal or above-normal weight.

45. b. As stated in the opening sentence of the fourth paragraph, bulimia patients may exercise obsessively. The other choices are not supported by the passage.

46. c. (A) and (B) are equally shaded. (A) and (B) both have 5 out of 9 boxes shaded and (C) only has 4 out of 9 boxes shaded.

47. k. 0.00102. When comparing decimals less than one, add zeros to the right of the last digit so that all the numbers have the same number of digits. After doing this, 0.00102 can be seen as having the smallest value since that is only 102 hundred-thousandths, whereas answer choice **j** is 12,000 hundred-thousandths.

48. a. 5. Let side $b = x$. Then side $a = 2x$, and side $b = \frac{1}{2}$ side c (which means $x = \frac{1}{2}c$ or $c = 2x$). The equation $a + b + c = 25$ can be written $x + 2x + 2x = 25$ by using substitution. $5x = 25$, so $x = 5$.

49. j. $\frac{1}{20}$. When calculating the probability that two events will both happen, multiply the individual probabilities together. The probability that Beth will be picked from the first hat is $\frac{1}{4}$ and the probability that Gillian will be picked from the second hat is $\frac{1}{5}$. The probability of both of these events happening is $\frac{1}{4} \times \frac{1}{5} = \frac{1}{20}$.

50. c. 15.
Sam $= s$. Ron $= \frac{1}{2} \times s$, Sam $= 3 \times$ Ted, which means Ted $= \frac{1}{3}(s)$.
Write an equation that models the sum of their ages: $s + \frac{1}{2}s + \frac{1}{3}s = 55$. The easiest way to solve this is to multiply by the entire equation by the least common multiple, which is 6, so that the fractions cancel out:
$6 \times s + 6 \times \frac{1}{2}s + 6 \times \frac{1}{3}s = 6 \times 55$
$6s + 3s + 2s = 330$, $11s = 330$, so $s = 30$.
Therefore Ron $= 15$.

51. f. irrational numbers. Irrational numbers make up a *different* set, not a *subset* of the rational numbers.

52. c. 93 million miles. When working with scientific notation with positive exponents, move the decimal that number of spaces to the right. In this case, the decimal gets moved 7 times, creating 6 zeros past the 93 which is equivalent to 93,000,000 or 93 million.

53. f. The mean and median scores are the same. The mean, or arithmetic average, is the sum of the numbers divided by the total number of data points. In this case the mean is 81. The median is the center number when the numbers are listed from least to greatest. In this case the median is 81. The mode is the most frequently occurring data point, which is 79 in this data set.

54. e. 12 feet by 18 feet. Set up a proportion with $\frac{\text{inches}}{\text{feet}}$: $\frac{0.5 \text{ inches}}{3 \text{ feet}} = \frac{2 \text{ inches}}{\text{width}}$, so width = 12. $\frac{0.5 \text{ inches}}{3 \text{ feet}} = \frac{3 \text{ inches}}{\text{length}}$, so length = 18.

55. h. 8. Remember that "of" means multiplication and "less than" means subtraction (but with "less than" you have to change the order of the terms). Set up and solve for y: $y = \frac{3}{4} \times 20 - 7$.

56. c. 320 square feet. Two walls will measure 9 feet wide by 8 feet tall, so their combined area is $2 \times 9 \times 8 = 144$. The other two walls will measure 11 feet long by 8 feet tall, so their combined area is $2 \times 11 \times 8 = 176$. $144 + 176 = 320$.

57. f. $6.00. To determine 80% of $7.50, turn 80% into a decimal and multiply it by $7.50. (Remember that "of" translates to multiplication in math, and that to turn a percentage into a decimal, you must move the decimal place two times to the left.) $(0.80)(\$7.50) = \$6.00$.

58. e. $50. The electricity cost for sales in 1999 was $750 and the electricity cost for customer service in 2000 was $700.

59. j. The engineering department showed a steady increase in the dollar amount of electricity used from 2000 to 2002.

60. b. 20%. Percent decrease or percent increase is always calculated by dividing the difference over the original. In this case the electricity usage from 1999 to 2000 must be divided by the original electricity usage in 1999: $\frac{1,000 - 800}{1,000} = 20\%$.

61. j. 13. The distance formula is $d = \sqrt{(x_2 - x_1)^2 + (y_2 - y_1)^2}$ $d = \sqrt{(6-1)^2 + (0-12)^2} = \sqrt{25 + 144} = \sqrt{169}$ $= \sqrt{169} = 13$.

62. c. 125 in³. $V = (\text{side})^3 = 5^3 = 125$ and volume is always written with an exponent of 3 attached to the units, which in this case is inches.

63. j. 13. $1.5x - 7 + 7 = 12.5 + 7$, $1.5x = 19.5$, $x = \frac{19.5}{1.5} = 13$.

64. d. $-8\frac{5}{7}$. Change the mixed fractions into improper fractions and change the "minus negative" to "plus": $-\frac{86}{7} + \frac{25}{7} = -\frac{61}{7} = 8\frac{5}{7}$.

65. h. $\frac{1}{2}$. Each number in this series is decreasing by $\frac{1}{8}$, so the next fraction would be $\frac{4}{8}$, which is equivalent to $\frac{1}{2}$.

66. b. $223.62. Multiply the number of people by the per-person cost to get the total cost: $(\$37.27)(6) = \$223.63$.

67. f. $6.50. Package B will cost $5.50 and package D will cost $5.50 also, so their combined cost will be $11. Package A will cost $3.00 and package C will cost $1.50, so their combined cost will be $4.50. The difference between $11 and $4.50 is $6.50.

68. d. $144,872. To find the total amount of money raised, multiply the average donation by the number of donors: $(\$796)(182) = \$144,872$.

69. g. 250 km. Pisa and Venice are approximately 2.5 units away from each other on the map, and 2.5×100 km = 250 km.

70. a. $2x^2 + 12x - 15$.
$5x^2 + 9x - 3 - 3(x^2 - x + 4)$
$5x^2 + 9x - 3 + -3(x^2 - x + 4)$ (first change the minus to plus negative)
$5x^2 + 9x - 3 - 3x^2 + 3x - 12$ (next distribute)
$5x^2 - 3x^2 + 9x + 3x - 3 - 12$ (group like terms)
$2x^2 + 12x - 15$ (combine like terms)

71. j. 13. $14^2 = 196$ and $13^2 = 169$, and since 169 is closer to 181 than 196, 13 is the best estimate.

72. e. 40 cm². Break this shape into a larger and smaller quadrilateral by extending down the center vertical line. The area of the larger quadrilateral is $5 \times 6 = 30$ cm² and the area of the smaller quadrilateral will be $5 \times 2 = 10$ cm². Their sum is 40 cm².

73. g. 5.25.
Point 1 is: $5\frac{1}{6}$ ft = 5.166
Point 2 is: 5.2 ft
Point 3 is: 5.25 ft

74. c. the number of people who are in the stadium before the football game ends. You do not know how many people left during the second half so you cannot calculate how many people remained in the stadium before the game ended.

75. f. The cylinder will overflow. You are given (Volume of a rectangular prism) = (Area of Base)(h), so the volume of the prism when it is filled 9 inches from the bottom is 16×9 = 144 in³. The maximum potential volume of the cylindrical container with a radius of 2 inches and a height of 11 inches is $V = (\pi r^2)(h) = 3.14 \times 4 \times 11 = 138.16$ in³. Since 144 in³ is greater than 138.16 in³, the rectangular box will contain more water than the cylinder can hold, so the cylinder will overflow.

76. a. $\frac{1}{7}$. When multiplying like bases, add the exponents: $7^{-3} \times 7^2 \times 7^4 \times 7^{-4} = 7^{(-3+2+4-4)} = 7^{-1} = \frac{1}{7}$ since a negative exponent gets moved the to bottom of a fraction and becomes positive.

77. g. $\frac{2(45)}{8} - 5\frac{1}{4}$.
The formula for the area of a trapezoid is $A = \frac{1}{2}h(b_1 + b_2)$. The height in the given trapezoid is 8, one of the bases is $5\frac{1}{4}$, and its area is 45, so these numbers can be plugged into the equation:
$A = \frac{1}{2}h(b_1 + b_2)$
$45 = \frac{1}{2}(8)(5\frac{1}{4} + b_2)$
Now multiply by 2 to get rid of the fraction:
$2(45) = 2\frac{1}{2}(8)(5\frac{1}{4} + b_2)$
$2(45) = 8(5\frac{1}{4} + b_2)$
$2(45) = 8(5\frac{1}{4}) + 8(b_2)$
$2(45) - 8(5\frac{1}{4}) = 8(b_2)$
$\frac{2(45) - 8(5\frac{1}{4})}{8} = b_2$
$\frac{2(45)}{8} - \frac{8(5\frac{1}{4})}{8} = b_2$
$\frac{2(45)}{8} - 5\frac{1}{4} = b_2$

78. a. $6(x^2)y + 2xy$. When multiplying like bases, add their exponents.

79. j. 246. Divide the total donation by the amount donated per house: $\$1{,}291.50 \div \$5.25 = 246$.

80. c. (–1,2). The coordinates of point P are (1,–3). Two units to the left of point P will give an x-coordinate of –1 and five units above it will give a y-coordinate of 2 for point l.

81. h. $9. Divide $576 by two to calculate how much Samantha earned each week. $576 ÷ 2 – $288. Next, divide $288 by 32 hours to compute her hourly rate of pay: $288 ÷ 32 = $9.

82. a. centimeters. Kilometers and miles are best used to measure long distances, as in travel by car. Yards and meters are common for measuring fabric or the height of buildings. Centimeters and inches would be appropriate for measuring books.

83. h. 5:24. In Derrick's sock drawer, there were 24 socks in total. Since there were 5 blue socks, the odds he pulled out a blue sock are 5:24.

84. a. 6 hours and 42 minutes. Set up a proportion modeling the number of pages Michael reads compared to minutes, beginning with a ratio modeling that for every 3 minutes, Michael reads 2 pages: $\frac{\text{minutes}}{\text{pages}} = \frac{3}{2}$ $= \frac{\text{\# minutes needed}}{268}$. Solving this ratio, Michael will need 402 minutes. 6 hours is 360 minutes and there are 42 minutes left over.

85. k. 480. Starting with Sunday, use the scale on the left hand side to estimate the number of tickets sold each day: $50 + 100 + 80 + 70 + 50 + 60 + 70 = 480$.

86. d. 8. 2, 4, 6, and 12 are all factors of 12 and divide it evenly, but 8 does not.

87. f. $7\frac{1}{2}$ inches. Set up a proportion modeling width to height: $\frac{\text{Width}}{\text{Height}} = \frac{8}{12} = \frac{5}{\text{new height}}$, so $60 = 8(\text{new height})$ and therefore the new height is $\frac{60}{8}$ inches which simplifies to $7\frac{1}{2}$ inches.

88. b. Jim makes it to Carnegie Hall and becomes a star.

89. f. 23. Set up two equations to model this problem. Let x = larger number and y = smaller number. The larger number x can be represented with $x = 6 + 3y$. Next, using their sum, you know that $x + y = 98$, which can be turned into $x = 98 - y$. Use substitution to put $98 - y$ into the first equation for x and then solve for y:
$98 - y = 6 + 3y$
$92 = 4y$, so $y = 23$.

90. e. 36 in². The perimeter of a square is $4 \times$ (side length). Since the perimeter is 24 inches, the side length must be 6 inches. Area of a square = (side length)², so the area of this square is 36 in.².

91. k. None of the above. The bottom, top, and left-facing surfaces must contain 6 unit squares—2 for the bottom, two top-facing and two left-facing. None of the answer choices have 6 unit squares for these faces.

92. d. $\frac{7}{24}$. $\frac{1}{3}$ of the $\frac{7}{8}$ of people that Jackie talked to think that recycling is important *and* buy things made from recycled materials: $\frac{1}{3} \times \frac{7}{8} = \frac{7}{24}$.

93. g. $\frac{325}{1,000} = \frac{x}{970,000}$. Start with a ratio that models "part" to "whole": $\frac{325}{1,000}$. Next, set it equal to another ratio that puts 970,000 in the "whole" category, leaving the "part" to be solved for.

94. e. $2 \times 2 \times 3 \times 5$. Begin by breaking 60 into its factors and continue breaking each factor down until only prime numbers are being used: $60 = 30 \times 2 = 6 \times 5 \times 2 = 3 \times 2 \times 5 \times 2$.

95. k. 42.67 cm². One-third of 24 is 8, so the height would be 8. One-third of 32 is $\frac{32}{3}$ which would be the base. Area of a triangle $= \frac{1}{2}(\text{base} \times \text{height}) = \frac{1}{2}(8 \times \frac{32}{3}) = 42.67$.

CHAPTER

5

▶ **PRACTICE TEST 3**

The *SHSAT Power Practice* tests will help you prepare for the high-stakes exams given to students applying for New York City's specialized high schools. Each practice test consists of sample questions like those you will find on the official SHSAT.

The 45-question verbal section and 50-question math section were developed by education experts. These tests will show you how much you know and what kinds of problems you still need to study. Mastering these practice tests will allow you to reach your highest potential on the real SHSAT.

PART I VERBAL

Scrambled Paragraphs

Paragraph 1

(q) (r) (s) (t) (u)
(q) (r) (s) (t) (u)
(q) (r) (s) (t) (u)
(q) (r) (s) (t) (u)
(q) (r) (s) (t) (u)

Paragraph 2

(q) (r) (s) (t) (u)
(q) (r) (s) (t) (u)
(q) (r) (s) (t) (u)
(q) (r) (s) (t) (u)
(q) (r) (s) (t) (u)

Paragraph 3

(q) (r) (s) (t) (u)
(q) (r) (s) (t) (u)
(q) (r) (s) (t) (u)
(q) (r) (s) (t) (u)
(q) (r) (s) (t) (u)

Paragraph 4

(q) (r) (s) (t) (u)
(q) (r) (s) (t) (u)
(q) (r) (s) (t) (u)
(q) (r) (s) (t) (u)
(q) (r) (s) (t) (u)

Paragraph 5

(q) (r) (s) (t) (u)
(q) (r) (s) (t) (u)
(q) (r) (s) (t) (u)
(q) (r) (s) (t) (u)
(q) (r) (s) (t) (u)

Logical Reasoning

6. (a) (b) (c) (d) (e)
7. (f) (g) (h) (j) (k)
8. (a) (b) (c) (d) (e)
9. (f) (g) (h) (j) (k)
10. (a) (b) (c) (d) (e)
11. (f) (g) (h) (j) (k)
12. (a) (b) (c) (d) (e)
13. (f) (g) (h) (j) (k)
14. (a) (b) (c) (d) (e)
15. (f) (g) (h) (j) (k)

Reading

16. (a) (b) (c) (d) (e)
17. (f) (g) (h) (j) (k)
18. (a) (b) (c) (d) (e)
19. (f) (g) (h) (j) (k)
20. (a) (b) (c) (d) (e)
21. (f) (g) (h) (j) (k)
22. (a) (b) (c) (d) (e)
23. (f) (g) (h) (j) (k)
24. (a) (b) (c) (d) (e)
25. (f) (g) (h) (j) (k)
26. (a) (b) (c) (d) (e)
27. (f) (g) (h) (j) (k)
28. (a) (b) (c) (d) (e)
29. (f) (g) (h) (j) (k)
30. (a) (b) (c) (d) (e)

31. (f) (g) (h) (j) (k)
32. (a) (b) (c) (d) (e)
33. (f) (g) (h) (j) (k)
34. (a) (b) (c) (d) (e)
35. (f) (g) (h) (j) (k)
36. (a) (b) (c) (d) (e)
37. (f) (g) (h) (j) (k)
38. (a) (b) (c) (d) (e)
39. (f) (g) (h) (j) (k)
40. (a) (b) (c) (d) (e)
41. (f) (g) (h) (j) (k)
42. (a) (b) (c) (d) (e)
43. (f) (g) (h) (j) (k)
44. (a) (b) (c) (d) (e)
45. (f) (g) (h) (j) (k)

PART II MATHEMATICS

46. (a) (b) (c) (d) (e)
47. (f) (g) (h) (j) (k)
48. (a) (b) (c) (d) (e)
49. (f) (g) (h) (j) (k)
50. (a) (b) (c) (d) (e)
51. (f) (g) (h) (j) (k)
52. (a) (b) (c) (d) (e)
53. (f) (g) (h) (j) (k)
54. (a) (b) (c) (d) (e)
55. (f) (g) (h) (j) (k)
56. (a) (b) (c) (d) (e)
57. (f) (g) (h) (j) (k)
58. (a) (b) (c) (d) (e)
59. (f) (g) (h) (j) (k)
60. (a) (b) (c) (d) (e)
61. (f) (g) (h) (j) (k)
62. (a) (b) (c) (d) (e)

63. (f) (g) (h) (j) (k)
64. (a) (b) (c) (d) (e)
65. (f) (g) (h) (j) (k)
66. (a) (b) (c) (d) (e)
67. (f) (g) (h) (j) (k)
68. (a) (b) (c) (d) (e)
69. (f) (g) (h) (j) (k)
70. (a) (b) (c) (d) (c)
71. (f) (g) (h) (j) (k)
72. (a) (b) (c) (d) (e)
73. (f) (g) (h) (j) (k)
74. (a) (b) (c) (d) (e)
75. (f) (g) (h) (j) (k)
76. (a) (b) (c) (d) (e)
77. (f) (g) (h) (j) (k)
78. (a) (b) (c) (d) (e)
79. (f) (g) (h) (j) (k)

80. (a) (b) (c) (d) (e)
81. (f) (g) (h) (j) (k)
82. (a) (b) (c) (d) (e)
83. (f) (g) (h) (j) (k)
84. (a) (b) (c) (d) (e)
85. (f) (g) (h) (j) (k)
86. (a) (b) (c) (d) (e)
87. (f) (g) (h) (j) (k)
88. (a) (b) (c) (d) (e)
89. (f) (g) (h) (j) (k)
90. (a) (b) (c) (d) (e)
91. (f) (g) (h) (j) (k)
92. (a) (b) (c) (d) (e)
93. (f) (g) (h) (j) (k)
94. (a) (b) (c) (d) (e)
95. (f) (g) (h) (j) (k)

Part 1—Verbal

The Verbal Test includes 45 questions in these three sections:

- Scrambled Paragraphs, 5 paragraphs (each counts double)
- Logical Reasoning, 10 questions, numbered 6–15
- Reading, 30 questions, numbered 16–45

Scrambled Paragraphs

This section tests your ability to organize a paragraph well. There are five paragraphs, presented in scrambled order. Your job is to put them in the best order to make a clear, coherent paragraph. Each correct answer counts double; these five paragraphs are worth 10 points out of the 50-point verbal test.

The first sentence in each paragraph is given. The remaining five sentences are listed in random order. Read each group of sentences carefully, and then decide on the best arrangement for them. Use the blanks at the left of each sentence to number these sentences from 1 to 5, showing the order they should be in.

Paragraph 1
Wouldn't you love to have a pet tarantula to startle and astonish your friends?

_____ **Q.** First, you should make sure that the type of pet you want to adopt will be legal.

_____ **R.** Once you are sure your pet would be legal, you must also consider whether you can provide for all of the exotic pet's needs.

_____ **S.** You must investigate the local state and national laws that prohibit or limit owning specific species.

_____ **T.** These exotic pets can be fascinating, but prospective owners must make careful evaluations before taking responsibility for one of these unusual companions.

_____ **U.** Perhaps you would rather amaze people by walking down the street with your sleek and adorable pet ferret.

Paragraph 2
Frostbite usually occurs when the outer parts of the body—fingers, toes, ears, or nose—are exposed to below-freezing air for a long time.

_____ **Q.** That means that you need to come inside *before* you get cold.

_____ **R.** Even if you're warmly dressed, you must also use common sense: Don't stay outside until your fingers or toes start feeling numb.

_____ **S.** To avoid this condition, dress properly before going out.

_____ **T.** If you cannot get to a doctor right away, warm the affected area slowly, using warm—not hot—water.

_____ **U.** However, if you are out in the cold too long and think you have frostbite, you should try to get medical attention as soon as possible.

Paragraph 3
Solar and lunar eclipses both occur from time to time.

_____ **Q.** A solar eclipse, however, can only be viewed from a zone that is about 200 miles wide and covers about one-half of a percent of Earth's total area.

_____ **R.** A lunar eclipse can be viewed from anywhere on the nighttime half of Earth.

_____ **S.** On the other hand, during a solar eclipse the moon passes between Earth and the sun.

_____ T. Of the two types of eclipses, the most common is the lunar eclipse, which occurs when a full moon passes through Earth's shadow.

_____ U. Another difference between the two types of eclipses is the area from which they can be seen.

Paragraph 4

The English-language premiere of Samuel Beckett's play *Waiting for Godot* took place in London in August 1955.

_____ Q. But director Peter Hall convinced *Godot*'s backers to refrain from closing the play, at least until the Sunday reviews were published.

_____ R. The play's first audience was stunned by the avant-garde work, which had only five characters and a minimal setting (one rock and one bare tree).

_____ S. Opening night critics and playgoers alike responded to *Godot* with bafflement and derision.

_____ T. Harold Hobson's review in the *Sunday Times* managed to save the play.

_____ U. Hobson had the vision to recognize *Godot* for what history has proven it to be—a revolutionary moment in theater.

Paragraph 5

In the summer of 2000, researchers in Oregon's Malheur National Forest stumbled upon what they believe to be the world's largest living organism.

_____ Q. In this way, *Armillaria ostoyae* steals water and carbohydrates from the tree, which causes the tree to rot out on the inside and die from lack of nutrients.

_____ R. This astounding find is a fungus that scientists know as *Armillaria ostoyae*, which is commonly called the *honey mushroom*.

_____ S. The underground structure of the fungus stretches across 2,200 acres and is still spreading.

_____ T. The rhizomorphs attach themselves the roots of some tree species, and then they send threadlike objects called *mycelia* into the roots.

_____ U. The fungus spreads by sending out yarnlike fingers called *rhizomorphs*.

Logical Reasoning

The questions in this section test your ability to reason well, that is, to figure out what the information you know can or can't possibly mean. Read the statements carefully, then choose the best answer based *only* on the information given. Note carefully the words used in each question. For example, one thing can be lar*ger* than another without being the lar*gest* in the group. In answering some of these questions, it may be useful to draw a rough diagram or make a list that gives real world values to the information.

6. Read the following statements:
1. Betty is older than Megan.
2. Polly is older than Betty.
3. Megan is older than Polly.

If the first two statements are true, then statement three is

a. true.
b. false.
c. partly true.
d. uncertain.
e. not possible to determine from the information given.

7. Jamie found a "buy a dozen, get half a dozen free" sale on long-stem roses. She walked out of the store carrying 36 long-stem roses.

Which of the following can we conclude is true?

 f. Jamie especially likes that store.
 g. Jamie paid for three dozen roses.
 h. Roses are Jamie's favorite flowers.
 j. Jamie always buys roses by the dozen.
 k. Jamie paid for two dozen roses.

8. Michael has a 6-pound puppy named Lucy, and Ilsa has a 35-pound puppy named Thurber. Michael and Ilsa are both very fond of their dogs.

Based on the statements above, which of the following must be true?

 a. Michael loves his puppy more than Ilsa loves hers.
 b. Lucy is smaller than Thurber.
 c. Thurber is older than Lucy.
 d. Thurber is smarter than Lucy.
 e. It cannot be determined from the information given.

Read the following, then answer Questions 9 through 11.

Five brothers—Randall, Samuel, Terrill, Ulysses, and Vernon—are each responsible for one housekeeping task—mopping, sweeping, laundry, vacuuming, or dusting. Each has one weekday when he performs his task and no more than one housekeeping task is performed per weekday.

- Vernon does not vacuum and does not do his task on Tuesday.
- Samuel does the dusting, and does not do it on Monday or Friday.

- The mopping is done on Thursday.
- Terrill does his task, which is vacuuming, on Wednesday.
- The laundry is done on Friday, and not by Ulysses.
- Randall does his task on Monday.

9. When does Samuel do the dusting?
 f. Monday
 g. Tuesday
 h. Wednesday
 j. Thursday
 k. Friday

10. What task does Vernon do?
 a. sweep
 b. dust
 c. vacuum
 d. mop
 e. laundry

11. Who does the sweeping?
 f. Randall
 g. Samuel
 h. Terrill
 j. Ulysses
 k. Vernon

12. Read the following statements:
 1. All spotted Gangles have long tails.
 2. Short-haired Gangles always have short tails.
 3. Long-tailed Gangles never have short hair.

If the first two statements are true, the third statement is

 a. true.
 b. false.
 c. probably false.
 d. partly true.
 e. not possible to determine from the information given.

13. According to research published in a recent issue of the *Princeton Review*, a student must receive a verbal score of at least 700 on his or her SAT to be considered for acceptance to Brown University. Laura is a second-year student at Brown University, majoring in comparative literature.

Which of the following can we conclude is true?

f. Laura did not spend very much time studying for the SAT.

g. Laura's verbal SAT score was approximately 550.

h. Laura's verbal SAT score was 700 or higher.

j. Laura does not like her major.

k. Laura's math SAT score was approximately 700.

14. The Bolshoi Theater is one of Moscow's oldest and most esteemed theaters. Last summer, Max visited the Bolshoi Theater and saw a brilliant ballet performance of *Swan Lake*.

Based on this information, we can conclude that:

a. The Bolshoi Theater is the center of Moscow's cultural life.

b. Many tourists visit the Bolshoi Theater every season.

c. Some of the most incredible performers of our time have been on the stage of the Bolshoi Theater.

d. Max had been to Moscow last summer.

e. It cannot be determined from the information given.

15. Ten new television shows appeared during the month of September. Five of the shows were sitcoms, three were hour-long dramas, and two were news-magazine shows. By January, only seven of these new shows were still on the air. Five of the shows that remained were sitcoms.

Based only on the information above, which of the following statements must be true?

f. Only one of the news-magazine shows remained on the air.

g. Only one of the hour-long dramas remained on the air.

h. At least one of the shows that was cancelled was an hour-long drama.

j. At least one of the shows that was cancelled was a news-magazine show.

k. Television viewers prefer sitcoms to hour-long dramas.

Reading Comprehension

Read each passage carefully and answer the questions that follow it. If necessary, you can reread the passage to be certain of your answers. Your answers must be based only on information that is actually in the passage.

Read the following passage, then answer Questions 16 through 20.

In 1997, the city of Moscow, capital of the newly independent nation of Russia, celebrated the 850th anniversary of its founding. In the more than eight centuries that Moscow has existed, it has been characterized by waves of new construction, the most recent one continuing into the present. Its architecture represents a hodgepodge of styles, as twelfth-century forms mingle with elegant estates from the times of the czars and functional structures that reflect the pragmatism of the Soviet era. As Moscow grows under a new system of government, there is concern that some of the city's architectural history will be lost.

Moscow has a history of chaotic periods that ended with the destruction of the largely wooden city and the building of the "new" city on top of the rubble of the old. The result is a layered city, with each tier holding information about a part of Russia's past. In some areas, archaeologists have reached the layer from 1147, the year of Moscow's founding. Among the findings from the various periods of Moscow's history are carved bones, metal tools, pottery, glass, jewelry, and crosses.

Russia has begun an enormous attempt to salvage and preserve as much of Moscow's past as possible. Realizing that, unless precautions are taken, new construction could destroy valuable historical structures and artifacts, the Department of Preservation of Historical Monuments is ensuring that this work is done in a manner that respects the past. Five thousand buildings have been designated as protected locations.

One example of the work done by the Department of Preservation is Manege Square. Throughout Moscow's past, this square has been a commercial district. In keeping with that history, the area will be developed as a modern shopping mall, complete with restaurants, theaters, and a parking garage. Before construction could begin, however, the site was excavated and a wealth of Russian history was uncovered.

Archaeologists working in Manege Square uncovered the commercial life of eight centuries and with it a profusion of historical artifacts. By excavating five meters deep, archaeologists were able to reconstruct the evolution of commercial Moscow. Among the finds are wooden street pavement from the time of Ivan the Terrible (sixteenth century), a

wide cobblestone road from the era of Peter the Great (early eighteenth century), street paving from the reign of Catherine the Great (mid to late eighteenth century), and a wealthy merchant's estate (nineteenth century). The citizens of the present are determined that history will not repeat itself, and that the past will be uncovered and celebrated rather than shrouded and forgotten.

16. Which of the following best tells what this passage is about?
    a. Unlike their predecessors, current citizens of Moscow are taking a respectful approach to modernizing their 850-year-old city, allowing new building while preserving important structures and artifacts from the past.
    b. The architecture of Moscow is a strange mix of styles, brought about by chaotic periods in its history during which new structures were built on top of old, or else older structures were simply destroyed.
    c. Today, approximately 5,000 active archeological sites in Moscow are being preserved by the Department of Preservation of Historical Monuments, and every day archeologists are making new and important discoveries.
    d. Because its architectural treasures are now being preserved, Moscow is faring much better than it did when it was part of the Soviet Union and under communist rule.
    e. Manege Square is a good example of what archaeologists and community members can do to satisfy preservation societies.

17. The passage states that Moscow's Manege Square has been and most likely will continue to be a
   f. residential area.
   g. community meeting place.
   h. business district.
   j. government center.
   k. cultural museum.

18. Which of the following can be inferred from the information in the second paragraph?
   a. Historically, the people of Moscow were more interested in modernization than in preservation.
   b. The Soviet government destroyed many old buildings, in keeping with an anti-czarist policy.
   c. There are very few 850-year-old cities in existence and fewer yet that preserve their past.
   d. Moscow has a history of invasions, with each new conqueror destroying the buildings of the previous regime.
   e. Russia officially became a nation in 1147.

19. Which of the following assumptions is most likely a viewpoint that would be expressed by the writer of this passage?
   f. Generally speaking, people should be more interested in building new structures than in saving old ones.
   g. Architectural history has little meaning to people struggling to form a new government.
   h. Progress and preservation are equally important principles of urban planning.
   j. Archaeologists and bureaucrats generally do not work well together.
   k. Architectural function is more important than architectural style

20. According to the passage, which of the following is true of archaeologists in Moscow?
   a. They have uncovered a number of significant items that will help them to understand Moscow's history.
   b. They operate under severe time constraints, as contractors wait to begin new buildings.
   c. There are not nearly enough archaeological teams to conduct all the possible research.
   d. They are primarily concerned with preserving the artifacts of modes of transportation.
   e. They are opposed to any more new construction in Manege Square.

Read the following passage, then answer Questions 21 through 24.

The skyline of St. Louis, Missouri, is fairly unremarkable, with one huge exception—the Gateway Arch that stands on the banks of the Mississippi as part of the Jefferson National Expansion Memorial. The arch is an amazing structure built to honor St. Louis's role as the gateway to the West.

In 1947, a group of interested citizens known as the Jefferson National Expansion Memorial Association held a nationwide competition to select a design for a new monument that would celebrate the growth of the United States. Other U.S. monuments are spires, statues, or imposing buildings, but the winner of this contest was a plan for a completely different type of structure. The man who submitted the winning design, Eero Saarinen, later became a famous architect. In designing the arch, Saarinen wanted to "create a monument which would have lasting significance and would be a landmark of our time."

The Gateway Arch is a masterpiece of engineering, a monument even taller than the Great Pyramid in Egypt. In its own way, the arch is at least as majestic as the Great Pyramid. The shape of the Gateway Arch is that of an inverted catenary curve, the same shape that a heavy chain will form if it is suspended between two points. The arch is covered with a sleek skin of stainless steel that often reflects dazzling bursts of sunlight. In a beautiful display of symmetry, the height of the arch is the same as the distance between the legs at ground level. The legs of the arch are equilateral triangles that decrease gracefully as the height increases. Inside the arch, there are structural reinforcements and a complex tram system to take visitors to the top.

Construction of the arch on the St. Louis waterfront finally began in February 1964. First, excavators dug 60 feet into the ground to lay the foundations for the arch. Then, using derricks, cranes, and other equipment, each section of the arch was hoisted into place and carefully welded together. Although the project was very dangerous, no one was killed during the construction. When the arch was finished in October 1965, more than 5,000 tons of steel and 38,000 tons of concrete had been used in the structure. The overall cost of the project was $13 million.

Today, the arch stands high over the surrounding community, an architectural masterpiece and a beautiful tribute to the explorers and pioneers who passed through St. Louis on their journeys westward. It is administered by the National Park Service and staffed with park rangers. In a single day, 6,400 visitors can travel to the top of the arch by means of the special tram system that operates inside the structure. Although the windows at the top are small because of structural pressures, the view is magnificent. On a clear day, people can see the horizon 30 miles away or gaze at closer landmarks from a unique perspective.

21. Which of the following best tells what this passage is about?
    a. how and why a special monument was built
    b. why a city needs a monument for its skyline
    c. how monuments serve to inspire mankind
    d. why people build monuments to their past
    e. In one day, six thousand visitors can visit the top of the arch.

22. Who chose Saarinen's design for the arch?
    f. a team of St. Louis architects
    g. the Jefferson National Expansion Memorial Association
    h. Eero Saarinen's engineering partners
    j. the National Park Service
    k. It was chosen by popular vote.

23. The author compares the arch to the Great Pyramid in Egypt in order to
    a. show readers that the arch was more difficult to construct.
    b. convince readers that the arch is more beautiful than the Great Pyramid.
    c. make readers aware of how much time it took to build the arch.
    d. prove that modern architects equal those in ancient times.
    e. give readers some idea of the height of the arch.

24. You can tell that the author probably considers the arch to be
    f. amusing.
    g. old-fashioned.
    h. impressive.
    j. silly.
    k. terrifying.

Read the following passage, then answer Questions 25 through 29.

[Thomas Builds-the-Fire is a Spokane Indian living on the Spokane Indian Reservation.]

So Thomas went home and tried to write their first song. He sat alone in his house with his bass guitar and waited for the song. He waited and waited. It's nearly impossible to write a song with a bass guitar, but Thomas didn't know that. He'd never written a song before.

"Please," Thomas prayed.

But the song would not come, so Thomas closed his eyes, tried to find a story with a soundtrack. He turned on the television and watched *The Sound of Music* on channel four. Julie Andrews put him to sleep for the sixty-seventh time, and neither story nor song came in his dreams. After he woke up, he paced around the room, stood on his porch, and listened to those faint voices that echoed all over the reservation. Everybody heard those voices, but nobody liked to talk about them. They were loudest at night, when Thomas tried to sleep, and he always thought they sounded like horses.

For hours, Thomas waited for the song. Then, hungry and tired, he opened his refrigerator for something to eat and discovered that he didn't have any food. So he closed the fridge and opened it again, but it was still empty. In a ceremony that he had practiced since his youth, he opened, closed, and opened the fridge again, expecting an immaculate conception of a jar of pickles. Thomas was hungry on a reservation where there are ninety-seven different ways to say fry bread. . . .

As his growling stomach provided the rhythm, Thomas sat again with his bass guitar, wrote the first song, and called it "Reservation Blues."

—Sherman Alexie, from *Reservation Blues* (1995)

25. Which of the following best tells what this passage is about?
  a. Thomas is having difficulty writing a song until his growling stomach provides him with a rhythm and a theme.
  b. Hungry and tired, Thomas opens his refrigerator but discovers that he doesn't have any food.
  c. After dinner, Thomas writes two new songs about life on the reservation.
  d. Thomas goes to sleep watching *The Sound of Music* on television.
  e. Listening to the faint voices that echo all over the reservation, Thomas finds it easy to come up with a new song.

26. Based on the passage, we can conclude that Thomas
  f. does not take good care of himself.
  g. is poor.
  h. has always wanted to be in a band.
  j. is waiting for someone to help him.
  k. watches too much television.

27. Thomas titles the song "Reservation Blues." Based on this passage, you can expect the song to be about
  a. the good times he's had on the reservation.
  b. how he and his friends started a band.
  c. fry bread.
  d. the sounds he hears at night on the reservation.
  e. the difficulties of living on a reservation.

28. Why does Thomas keep opening and closing the refrigerator?
  f. He keeps hoping food will magically appear.
  g. He can't believe that the refrigerator is empty.
  h. He is angry and wants the door to break off.
  j. He likes the noise the door makes.
  k. He is bored.

29. The narrator tells us that "Thomas was hungry on a reservation where there are ninety-seven ways to say fry bread." What is the purpose of this sentence?
   a. to show us how important fry bread is to the language
   b. to show us how hungry Thomas was
   c. to make us want to try fry bread
   d. to show us the irony of the situation
   e. to show us how Thomas was inspired

Read the following passage, then answer Questions 30 through 34.

Consider the lowly alkaline battery. Compared to laptop computers, camcorders, and cell phones, this ordinary battery doesn't seem very important. But without it, none of these portable modern conveniences would be possible.

Although batteries have been available to the public since the turn of the twentieth century, until 1959 they were weak and provided only enough energy to power a dimly lit flashlight for a very short time. It was the invention of the alkaline battery that opened the door to the invention of laptop computers, calculators, and portable CD players—not to mention the portable computer games—that we take for granted today.

In the 1950s, there was a boom in the invention and marketing of household conveniences. During that decade, refrigerators, washers and dryers, and television sets—all designed to make life easier—became common fixtures in American homes. However, all those new inventions had to be plugged into the wall, which limited their use. Researchers decided it was time to improve the battery and allow for the development of portable appliances, such as the transistor radio.

All batteries work essentially the same way. Each has a cathode, or positive electrode, and an anode, or negative electrode. These are placed in a material known as an electrolyte that passes electrons between the cathode and the anode. This action is what creates electricity. However, as the electrons pass from the anode to the cathode, the capacity of the battery decreases and eventually the battery dies.

The alkaline battery changed all that. By using alkaline for the electrolyte, a cathode made of manganese dioxide and carbon, and an anode made of zinc, scientists developed a battery with more power and longer life. It wasn't until batteries had the capabilities of the alkaline battery that a whole array of portable devices could be marketed. Today, most homes contain about 18 different devices that use batteries. Research suggests that, in the future, batteries will be even more powerful but much smaller. A battery as small as the one in a wristwatch will have enough power to run a cell phone.

30. Which of the following best tells what this passage is about?
   a. Scientists have strengthened the previously weak alkaline battery.
   b. Battery-driven devices are replacing common household fixtures.
   c. The alkaline battery has affected the development and marketing of portable devices.
   d. This is how alkaline batteries are produced and manufactured.
   e. American homes can reduce the amount of electricity they use.

31. As compared with batteries made after 1959, batteries made before 1959
  f. were weaker and had a shorter life span.
  g. had more power but a shorter life span.
  h. were created for use in household appliances.
  j. were made without cathodes and anodes.
  k. did not use an electrolyte.

32. The passage suggests that alkaline batteries create electricity by
  a. passing positive and negative electrodes through a zinc electrolyte.
  b. passing electrons between manganese dioxide and carbon.
  c. exchanging the negative cathode with a positive one.
  d. transforming negative electrodes into anodes.
  e. passing electrons between the cathode and the anode.

33. What made researchers in the late 1950s decide that batteries needed to be improved?
  f. Batteries were weaker and had a shorter life span.
  g. Batteries had more power but a shorter life span.
  h. Batteries were created for use in household appliances.
  j. Batteries were made without cathodes and anodes.
  k. Batteries did not use an electrolyte.

34. The last paragraph states that batteries will be smaller and more powerful in the future. This suggests that
  a. research for improving batteries is currently under way.
  b. batteries will replace electricity as a primary power source.
  c. household appliances will also be smaller.
  d. portable appliances will eventually become obsolete.
  e. batteries will require increased amounts of alkaline.

Read the following passage, then answer questions 35 through 38.

He laughs and asks, "Do I have a silly laugh?"

"No," I say, "no." But he does. Jorge does have a silly laugh. Faw faw fawwww. Faw faw fawwww. With squeaks. You'd actually get up and change seats in a movie theater if you were sitting behind somebody who had this laugh, because how could you watch the movie with that going on?

"People have told me I have a silly laugh."

"Oh no," I say, "no." My words are specific, chosen. Each no is its own precise lie. One means "No, you don't have a silly laugh." The next means "No, you shouldn't listen to your friends." Another means "No, I wouldn't say that since, after all, you're doing me a favor by driving me to school." I could lie more explicitly, but the words just come out "no, no, no."

"I'm glad your sister got sick," he says, and wham!

. . . After a moment of embarrassed silence, Jorge realizes what he has just said and nervously offers an explanation.

"I mean, I'm not glad she got sick. When I said I was glad she got sick, I didn't mean it. What I meant was that I'm glad she couldn't drive you to school. I mean, I'm not glad she couldn't, but I'm glad that I get to, you know? Because I've been wanting to meet you ever since you moved in."

I nod. I should say something like "yes, me too," or "I've wanted to get to know you too, Jorge," but those lies catch in my throat.

The silence in the car becomes thicker as we slow to a halt. Outside the window, a worker is redirecting traffic onto the shoulder while a construction crew prepares to dig up a big chunk of the highway. Inside the car, Jorge works to fill the silence with more hot air.

"I mean, we are neighbors now, so we should know more about each other, right? Good neighbors make good fencers," Jorge says and laughs again. Faw faw fawwww.
—Eric Reade, "Discomfort and Joy"

35. Which of the following best tells what this passage is about?
   a. Old friends, Jorge and the narrator are riding to school together as usual.
   b. The narrator expects to become close friends with his neighbor Jorge.
   c. The narrator is annoyed with Jorge's driving and will not ride with him again.
   d. It's clear that the narrator and Jorge take turns driving each other to school.
   e. Jorge is driving the narrator to school, but neither of them is very comfortable.

36. What event probably occurred before the beginning of this excerpt?
   f. Jorge drove the narrator to school for several days.
   g. The narrator's sister recovered from her illness.
   h. Jorge asked whether he has a silly laugh.
   j. The narrator asked Jorge for a ride to school.
   k. The narrator noticed road construction on the way to school.

37. Why does the narrator lie to Jorge in the opening lines of this excerpt? The narrator
   a. wants to avoid the awkwardness that would come from telling the truth.
   b. hopes to become close friends with Jorge.
   c. is a compulsive liar who is unable to distinguish between lies and the truth.
   d. plans to tell Jorge the truth about his laugh when they reach the school.
   e. wants to punish Jorge for his poor jokes.

38. The title of this story is "Discomfort and Joy." What does this title suggest will happen between the narrator and Jorge?
   f. The tension between them will probably explode into violence.
   g. The tension between them be resolved, leading to a better relationship between the narrator and Jorge.
   h. The tension between them will continue long after the ride in the car.
   j. The tension between them will increase steadily as the car gets closer to the school.
   k. The tension between them will turn out to have been in the narrator's imagination all the time.

Read the following passage, then answer Questions 39 through 45.

New research suggests that animals have a much higher level of brainpower than previously thought. But even if animals do have real intelligence, how can scientists measure it?

Before defining animals' intelligence, scientists defined what is not intelligence. *Instinct* is not intelligence. It is a skill programmed into an animal's brain by its genetic heritage. *Rote conditioning* is also not intelligence. Tricks can be learned by repetition, but no real thinking is involved. *Cuing*, in which animals learn to do or not to do certain things by following outside signals, does not demonstrate intelligence.

Scientists believe that insight, the ability to use tools, and communication using human language are all effective measures of the mental ability of animals.

When judging animal intelligence, scientists look for insight, which they define as a flash of sudden understanding. When a young female gorilla could not reach fruit from a tree, she noticed crates scattered about the lawn near the tree. She piled the crates into a pyramid, then climbed on them to reach her reward. The gorilla's insight allowed her to solve a new problem without trial and error.

The ability to use tools is also an important sign of intelligence. Crows use sticks to pry peanuts out of cracks. A crow exhibits intelligence by showing it has learned what a stick can do. Likewise, otters use rocks to crack open crab shells in order to get at the meat. Chimpanzees have been known to use sticks and stalks in a series of complex moves to get at a favorite snack—termites.

To make and use a termite tool, a chimp first selects just the right stalk or twig. The animal trims and shapes the stick, then finds the entrance to a termite mound. While inserting the stick carefully into the entrance, the chimpanzee turns it skillfully to fit the inner tunnels. Then the chimp attracts the insects by shaking the twig. Careful not to scrape off any termites, the chimp pulls the tool out. Finally, the chimp uses its own lips to skim the termites into its mouth.

Many animals have learned to communicate using human language. Some primates have learned hundreds of words in sign language. One chimpanzee can recognize and correctly use more than 250 abstract symbols on a keyboard. These symbols represent human words.

An amazing parrot can distinguish five objects of two different types. He can understand the difference between the number, color, and kind of object. The ability to classify is a basic thinking skill. The parrot also seems to use language to express his needs and emotions. When ill and taken to the animal hospital for his first overnight stay, this parrot turned to go. "Come here!" he cried to a scientist who works with him. "I love you. I'm sorry. Wanna go back?"

The research on animal intelligence raises important questions. If animals are smarter than once thought, would that change the way humans interact with them? Would humans stop hunting them for sport or survival? Would animals still be used for food, clothing, or medical experimentation? Finding the answer to these tough questions makes a difficult puzzle even for a large-brained, problem-solving species like our own.

39. Which of the following best tells what this passage is about?
   a. Some animals seem to instinctively display signs of high intelligence.
   b. An amazing parrot can classify materials; classifying is a basic thinking skill.
   c. Chimpanzees have been known get termites to eat by using a complex series of actions.
   d. Some animals can use human language to communicate.
   e. New research shows more signs of intelligence in animals than was previously thought.

40. Crows use sticks to pry peanuts out of cracks. Which of the following is the kind of intelligence or conditioning the situation describes?
  f. rote learning
  g. tools
  h. communication
  j. instinct
  k. cueing

41. According to the passage, animal intelligence has been indicated by
  a. success in the rote conditioning of many animals.
  b. observation of animals in the wild.
  c. observation of animals in zoos.
  d. a new study of instinctive animal behavior.
  e. observation of animals using tools.

42. The concluding paragraph of this passage implies which of the following?
  f. There is no definitive line between those animals with intelligence and those without.
  g. Animals are being given opportunities to display their intelligence.
  h. Research showing higher animal intelligence may fuel debate on ethics and cruelty.
  j. Animals are capable of untrained thought well beyond mere instinct.
  k. Animals are capable of acting intelligently based on instinct.

43. According to the passage, which of the following is true about animals communicating through the use of human language?
  a. Parrots can imitate or repeat a sound.
  b Dolphins click and whistle.
  c. Crows screech warnings to other crows.
  d. Chimpanzees and gorillas have been trained to use sign language.
  e. Animals seem to communicate with each other.

44. According to the passage, what conclusion can be reached about the chimpanzee's ability to use a tool to get food?
  f. It illustrates instinct because he is able to get his food and eat it.
  g. It illustrates intelligence because termites are protein-packed.
  h. It illustrates instinct because he faced a difficult task and accomplished it.
  j. It illustrates rote learning because he stored knowledge away and called it up at the right time.
  k. It illustrates high intelligence because it is a complex task that calls upon stored knowledge.

45. Which of the following is NOT a sign of animal intelligence?
  a. shows insight
  b. follows cues
  c. uses tools
  d. makes a plan
  e. classifies objects

Part 2—Math

The SHSAT Grade 8 Math Test includes 50 questions covering content in the following areas:

- basic math
- percentages, fractions, decimals, averages
- pre-algebra
- algebra
- substitution
- factoring
- coordinate graphing
- geometric principles
- logic
- word problems

Solve each problem and select the best answer from the choices given. It is important to keep in mind that:

- Formulas and definitions of mathematical terms and symbols are not provided.
- Diagrams other than graphs are not necessarily drawn to scale. Do not assume any relationship in a diagram unless it is specifically stated or can be figured out from the information given.
- A diagram is in one plane unless the problem specifically states that it is not.
- Graphs are drawn to scale. Unless stated otherwise, you can assume relationships according to appearance. For example, (on a graph) lines that appear to be parallel can be assumed to be parallel; likewise for concurrent lines, straight lines, collinear points, right angles, and so on.
- You will need to reduce all fractions to lowest terms.

46. What number is 3 more than 20% of 100?
  a. 17
  b. 23
  c. 28
  d. 60
  e. 123

47. If $\frac{1}{16} = \frac{x}{54}$, what is x?
  f. 3.375
  g. 3.75
  h. 3.5
  j. 4
  k. 4.5

48. Karl is four times as old as Pam, who is one-third as old as Jackie. If Jackie is 18, what is the sum of their ages?
  a. 5
  b. 8
  c. 10
  d. 15
  e. 48

49. What is the probability that the spinner will land on the number 1?

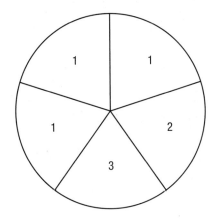

  f. $\frac{2}{5}$
  g. $\frac{2}{3}$
  h. $\frac{1}{3}$
  j. $\frac{3}{5}$
  k. $\frac{3}{4}$

50. Let m, n, and p represent real numbers with $m > 0$ and $n > p$. Which statement is NOT true?
  a. $m + n > m + p$
  b. $\frac{n}{m} > \frac{p}{m}$
  c. $m - n > m - p$
  d. $p - m > n - m$
  e. $nm > p(-m)$

51.

21 ft

Living Room

12 ft

Bernice is planning to carpet her living room. She has selected carpeting that costs $12 a square yard. What is the total cost for carpeting Bernice's living room?

- **f.** $3,024
- **g.** $1,008
- **h.** $336
- **j.** $252
- **k.** $224

52. A refrigerator has dimensions 2 feet by 3 feet by 5 feet. What is the volume of the refrigerator?

- **a.** 10 cubic feet
- **b.** 25 cubic feet
- **c.** 30 cubic feet
- **d.** 45 cubic feet
- **e.** 62 cubic feet

53. The product of 16 and one-half a number is 136. Find the number.

- **f.** 84
- **g.** 62
- **h.** 17
- **j.** 76
- **k.** 16

54. White flour and whole wheat flour are mixed together in a ratio of 5 parts white flour to 1 part whole wheat flour. How many pounds of white flour are in 48 pounds of this mixture?

- **a.** 8 pounds
- **b.** 9.6 pounds
- **c.** 24.4 pounds
- **d.** 40 pounds
- **e.** 42 pounds

55. Look at this series: 9, 12, 11, 14, 13, 16, 15, . . . What two numbers should come next?

- **f.** 14, 13
- **g.** 8, 21
- **h.** 17, 16
- **j.** 14, 17
- **k.** 18, 17

56. After three days, some hikers discover that they have used $\frac{2}{5}$ of their supplies. At this rate, how many more days can they go forward before they have to turn around?

- **a.** 0.75 days
- **b.** 1.5 days
- **c.** 3.75 days
- **d.** 4 days
- **e.** 6 days

57. A family's gas and electricity bill averages $80 a month for seven months of the year and $20 a month for the rest of the year. If the family's bills were averaged over the entire year, what would the monthly bill be?

- **f.** $42.50
- **g.** $45
- **h.** $50
- **j.** $55
- **k.** $60

58. What is the value of *x* in the figure below?

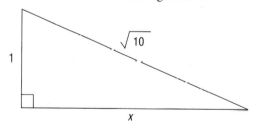

- **a.** 3
- **b.** 4
- **c.** 5
- **d.** 6
- **e.** 9

59. A large cube is composed of smaller cubes and measures 4 cubes long by 4 cubes wide by 4 cubes high. The outside of the large cube is painted blue. After the paint dries, the large cube is broken apart into the smaller cubes. How many of the smaller cubes will not be painted blue on any of their sides?

 f. 8
 g. 12
 h. 16
 j. 32
 k. 64

60. All pies at the baked goods booth were cut into eighths. When Justina Hartley's shift in the booth began, there were eight and one-fourth apple pies waiting to be sold. At the end of her shift there were only five pieces of apple pie left. How many pieces of apple pie were sold during Justina's shift?

 a. 60
 b. 61
 c. 62
 d. 63
 e. 64

61. The scale on an architectural floor plan is 1 inch:12 feet. The length of a hallway in the floor plan is 1.75 inches. What is the actual length, in feet, of the hallway?

 f. 15 feet
 g. 18 feet
 h. 21 feet
 j. 24 feet
 k. 28 feet

62. Solve the following equation for x: $\frac{3}{4}x + 7 = -14$

 a. 28
 b. −15.75
 c. −22.666 . . .
 d. −28
 e. −9.333 . . .

63. Which number is the result of evaluating the expression $(-5)^2 \times (2)^3$?

 f. −200
 g. −60
 h. 33
 j. 100
 k. 200

64. Find the missing angle measure.

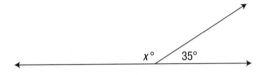

 a. 145°
 b. 45°
 c. 65°
 d. 135°
 e. 125°

65. Examine (A), (B), and (C) and find the best answer.

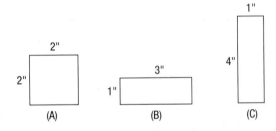

 f. The area of (C) is less than the area of (B).
 g. The area of (A) is equal to the area of (B).
 h. The area of (A) is equal to the area of (C).
 j. The area of (B) is equal to the area of (C).
 k. The area of (A) is less than the area of (B).

66. A 600-page book is 1.5 inches thick. What is the thickness of each page?

 a. 0.0010 inches
 b. 0.0030 inches
 c. 0.0025 inches
 d. 0.0600 inches
 e. 0.0750 inches

67. A horse is tied to a post with a twenty-foot rope. What is the longest path that the horse can walk?

   f. 20 feet

   g. 40 feet

   h. 20π feet

   j. 40π feet

   k. 400π feet

68. While driving home from work, Sally runs over a nail, causing a tire to start leaking. She estimates that her tire is leaking 1 psi every 20 seconds. Assuming that her tire leaks at a constant rate and her initial tire pressure was 36 psi, how long will it take her tire to completely deflate?

   a. 1.8 minutes

   b. 3.6 minutes

   c. 12 minutes

   d. 16.5 minutes

   e. 18 minutes

69. What value of b will make $\sqrt{144-b}$ a rational expression?

   f. 120

   g. 142

   h. 23

   j. 19

   k. 146

Use the following chart to answer Questions 70 and 71.

Food Item	Price
Hot dogs	1—$2.25 3—$6.00
Hamburgers	1—$3.50 3—$10.00
French fries	small—$2.00 jumbo—$3.00
Soda	8 oz—$1.25 12 oz—$1.50 20 oz—$2.00

70. Which expression shows the total cost for purchasing two hot dogs, a jumbo size order of french fries, and an 8-oz cup of soda?

   a. $2(3.5) + 3 + 1.5$

   b. $2(2.25) + 3 + 1.25$

   c. $2(2.25) + 2 + 1.5$

   d. $6 + 2(2) + 1.25$

   e. $7 + 2 + 1.35$

71. Frank wants to buy four hamburgers, two jumbo size fries, and three 12-oz sodas. What is the least amount he can spend to buy these items?

   f. $24.00

   g. $25.50

   h. $27.50

   j. $30.00

   k. $31.50

72. Which ordered pair best represents the coordinates of the point plotted in the following coordinate plane?

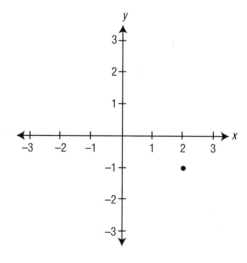

   a. (2,–1)

   b. (–1,2)

   c. (2,1)

   d. (–2,–1)

   e. (–2,1)

73. Multiply: $(x + 7)(x - 4)$

  f. $x^2 + 3x - 28$
  g. $x^2 - 28$
  h. $x^2 - 3x - 28$
  j. $x^2 + 3x + 3$
  k. $x^2 - 3x + 3$

74. Which is the *best* unit of measurement to measure the amount of water in a swimming pool?

  a. ounces
  b. cups
  c. feet
  d. gallons
  e. milliliters

75. For every dollar Kyra saves, her employer contributes a dime to her savings, with a maximum employer contribution of $10 per month. If Kyra saves $60 in January, $130 in March, and $70 in April, how much will she have in savings at the end of that time?

  f. $270
  g. $283
  h. $286
  j. $290
  k. $260

76. A rain barrel contained 4 gallons of water just before a thunderstorm. It rained steadily for 8 hours, filling the barrel at a rate of 6 gallons per day. How many gallons of water did the barrel have after the thunderstorm?

  a. 4 gallons
  b. 6 gallons
  c. 7 gallons
  d. 8 gallons
  e. 9 gallons

77. Find the radius of a circle with a circumference of 36π.

  f. 6π
  g. 18π
  h. 9
  j. 9π
  k. 18

78. The two boxes have the same volume ($V = l \times w \times h$). Find the missing side on box B.

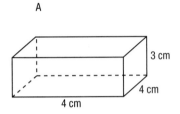
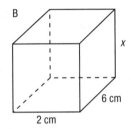

  a. 3 cm
  b. 4 cm
  c. 4.5 cm
  d. 5 cm
  e. 6 cm

79. The bar graph shows the DVD sales at a local video store over a period of five days. Which statement is true about the data displayed on the bar graph?

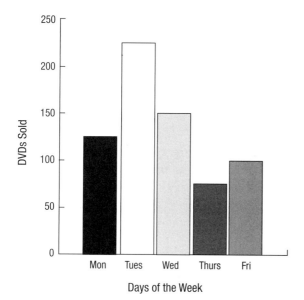

Video Store Sales

f. Fewer DVDs were sold on the last three days than on the first two days.

g. The number of DVDs sold increased every day.

h. The smallest number of DVDs was sold on Friday.

j. On Monday, 225 DVDs were sold.

k. There were a total of 665 DVDs sold on all five days.

80. Choose the statement that is the inverse of "If it rains, then I will take an umbrella."

a. If I take an umbrella, then it rains.

b. If I do not take an umbrella, then it does not rain.

c. If it rains, then I do not take an umbrella.

d. If it does not rain, then I do not take an umbrella.

e. If I do not take an umbrella, then it will not rain.

81. Find the value of a when $b = 6$.

$$12a + \frac{b^2}{4} = 93$$

f. 5

g. 6

h. 7

j. 8

k. 9

82. The table shows how many books a bookstore sold in 6 months. What is the most accurate estimate of how many books were sold from the beginning of July through the end of December?

Book Sales

Month	Number of Books Sold
July	3,377
August	8,499
September	7,726
October	5,622
November	7,124
December	11,294

a. 25,000

b. 29,000

c. 32,000

d. 39,000

e. 43,000

83. $|(2x^3y)^3|(4x^2y^2)$ is equivalent to which of the following?

f. $32x^{11}y^5$

g. $8x^{11}y^5$

h. $32x^{18}y^6$

j. $24x^{18}y^6$

k. $8x^{18}y^6$

84. What is the square root of $\frac{64}{25}$?

   a. $\frac{5}{8}$

   b. $\frac{6}{5}$

   c. $\frac{8}{5}$

   d. $2\frac{14}{25}$

   e. $6\frac{346}{625}$

85. The table shows the colors of replacement parts for Pocket PCs. The total number of parts shipped is 1,650.

BOXED SET OF REPLACEMENT PARTS	
Part Color	Number of Pieces
Green	430
Red	425
Blue	
Yellow	345
TOTAL	1,650

If a person randomly grabs a part out of the box, what is the probability that the part will be blue?

   f. $\frac{1}{4}$

   g. $\frac{5}{12}$

   h. $\frac{1}{9}$

   j. $\frac{1}{12}$

   k. $\frac{3}{11}$

86. A garden hose fills a 50-gallon drum with water in 5 hours. How long will it take the hose to fill a 72-gallon drum?

   a. 3 hours, 12 minutes

   b. 3.12 hours

   c. 7.12 hours

   d. 7 hours, 12 minutes

   e. 7 hours

87. Three apples and twice as many oranges add up to one-half the number of cherries in a fruit basket. How many cherries are there?

   f. 11

   g. 16

   h. 18

   j. 21

   k. 24

88. Swimming Pool World made a donation to a children's hospital. They pledged 3.2% of their sales for the second week of May. Below is their sales chart for May.

May Sales	
Week 1	$5,895
Week 2	$73,021
Week 3	$54,702
Week 4	$67,891

How much did Swimming Pool World donate to the children's hospital?

   a. $2,336.67

   b. $3,651.05

   c. $18,573.81

   d. $23,366.72

   e. $36,510.50

89. The high temperatures for the first six days of a week are shown in the table. What was the median temperature from Sunday to Friday?

High Temperatures	
Sunday	60°
Monday	62°
Tuesday	64°
Wednesday	61°
Thursday	65°
Friday	60°

f. 60°
g. 61.5°
h. 62°
j. 62.5°
k. 63°

90. The wheel of a bicycle has a diameter of 25 inches. How many turns does the wheel make in traveling 1 mile (5,280 feet)? Round your answer to the nearest full turn. The formulas for the area and circumference of a circle are as follows:

$A = \pi r^2$
$C = \pi d$
Use 3.14 for π.

a. 211
b. 79
c. 576
d. 807
e. 491

91. The diagram shows a pattern for making a cardboard cereal box. The box will be 11 inches high, 8 inches wide, and 2 inches deep. How much cardboard will be needed to make the box?

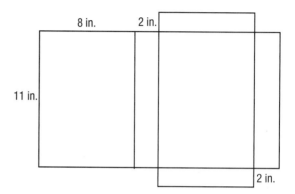

f. 110 in.^{2}
g. 176 in.^{2}
h. 194 in.^{2}
j. 220 in.^{2}
k. 252 in.^{2}

92. Jeremy purchased 6 cans of tomatoes for $5.34. At this rate, how much will he pay for 11 cans of tomatoes?
a. $10.68
b. $10.24
c. $9.79
d. $9.90
e. $11.00

93. Which of the following is NOT a prime number?
f. 1
g. 2
h. 67
j. 149
k. 401

94. The Ramirez family visited Canada. The rate of currency during their stay in Canada was 1.35 Canadian dollars per U.S. dollar. On their first day in Canada, Mr. and Mrs. Ramirez exchanged $500 U.S. for Canadian dollars. How many Canadian dollars did they get in the exchange?
a. $765 Canadian
b. $675 Canadian
c. $501.35 Canadian
d. $635 Canadian
e. $175 Canadian

95. What is the smallest number that is both a multiple of 6 and a multiple of 15?
f. 3
g. 21
h. 30
j. 45
k. 90

Answers

Paragraph 1 (U, T, Q, S, R)
The given sentence about a pet tarantula is logically followed by the alternative suggestion of a pet ferret in sentence **U**. The rest of the sentences contain cautions about adopting exotic pets. In sentence **T**, the words "These exotic pets" indicate that **T** follows the specific mentions of exotic pets, and **T** also makes a transition between the exotic pets and the specific cautions. Sentence **Q**, which states that it comes first, cautions about making sure your pet will be legal. Sentence **S** tells you how to do that. Sentence **R** states what you must consider after "you are sure your pet would be legal," so it follows **Q** and **S**.

Paragraph 2 (S, R, Q, U, T)
Sentence **S** directly follows the information in the given sentence. Sentence **R** refers back to **S**, so it comes next. The words "That means" in sentence **Q** indicate that it comes after some other statement, so it follows **R**. The words "However, if you are out in the cold too long" in sentence **U** show that it comes after the advice about not staying out in cold. Sentence **T** tells what to do if you can't follow the advice in sentence **U** (get to a doctor).

Paragraph 3 (T, S, U, R, Q)
Sentence **T** refers to the two types of eclipses mentioned in the given sentence. The words "On the other hand" in sentence **S** indicate that it follows some other statement, so it comes after **T**. Sentence **U** describes the viewing area as "Another difference." That means that **U** must come after the first difference (**T** and **S**), and **R** and **Q** must follow **T**.

Paragraph 4 (R, S, Q, T, U)

Sentence **R** refers to the audience at the premiere mentioned in the given sentence. Sentence **S** refers to the response of that audience, including critics, to the play after they had seen it, so **S** follows **R**. The remaining sentences concern the survival of the play. Sentence **Q** is about waiting for the Sunday reviews, and sentence **T** is about when Hobson's Sunday review was published. Sentence **U** is about the result of Hobson's review, so **Q**, **R**, and **T** must be placed in chronological order.

Paragraph 5 (R, S, U, T , Q)

In sentence **R**, "This astounding find" logically refers to the discovery mentioned in the given sentence, and **R** first describes the find as a fungus. Sentence **S** gives the size of the fungus and says that it is "still spreading." Sentence **U** tells how the fungus spreads, and first introduces rhizomorphs. Since sentence **T** describes how the rhizomorphs attach to tree roots, **T** must follow **U**. Sentence **Q** gives the result of the attachment to tree roots, so it comes after **T**.

6. b. The first two statements are true, so Megan is the youngest of the three. The third statement, therefore, is false.

7. k. Based on the information available, we can conclude that Jamie paid for two dozen roses, since she walked out carrying 36 roses (3 dozen). We know that the "buy a dozen, get half a dozen free" deal would allow her to have a dozen roses (two half dozens) for free, if she bought two dozen, so choice **g** is incorrect. Choices **f**, **h**, and **j** may be true, but we do not have enough information to reach those conclusions.

8. b. Based on the two statements, the only answer choice that we can conclude is correct is that Lucy is smaller than Thurber, because we are told Lucy is a 6-pound puppy, and Thurber is a 35-pound puppy. There is no point of comparison to measure the love of Michael and Ilsa for their puppies (choice **a**). Choices **c** and **d** could be true, but these statements are unsupported by the information.

9. g. The best way to solve this problem is to create a chart. Write down what you know immediately from the given information. Your chart should look something like this:

DAY	TASK	PERSON
M	? not dusting	Randall's job
T	?	?
W	vacuum	Terrill
Th	mop	?
F	laundry not dusting	? not Ulysses

Then, determine the missing parts of the chart by a process of elimination. Dusting and Samuel must be filled in on one day, and there's only one day when those two blanks are available. If Ulysses doesn't work on Friday, there's only one time left when he could work: Thursday. That just leaves Monday available for sweeping and Friday available for Vernon to work. Your finished chart should look something like this:

DAY	TASK	PERSON
M	sweep	Randall
T	dust	Samuel
W	vacuum	Terrill
Th	mop	Ulysses
F	laundry	Vernon

10. e. Refer to the chart you made for Question 9 and you will see that Vernon does the laundry.

11. f. Refer to the chart you made for Question 9 and you will see that the job of sweeping must belong to Randall.

12. a. The second statement tells us that if a Gangle has short hair, it *always* has a short tail. Therefore, a Gangle with a long tail can't possibly have short hair.

13. h. Based only on the information given, we do not know whether Laura does not like her major (choice **j**) or how much time she spent studying for the SAT (choice **f**). The passage tells us nothing about Laura's math score (choice **k**). We do know that Laura's score could not have been approximately 550 (choice **g**), because the first statement says that a student must receive a verbal score of at least 700 in order to be considered for acceptance to Brown University, and we are also told that Laura is a student at the university. Therefore, the only answer that is correct is choice **h**, Laura's score was 700 or higher.

14. d. Although **a**, **b**, and **c** may be true, there is nothing in the information provided to support those conclusions. The only thing known for certain is that Max, since he had been to the Bolshoi Theater, had visited Moscow last summer (choice **d**).

15. h. If there were seven shows left, and five were sitcoms, this means that only two of the shows could possibly be dramas. Since there were originally three new hour-long dramas, at least one of the dramas was cancelled. Choices **f**, **g**, and **j** might be true, but there is not enough evidence to indicate this as fact. The fact that all the sitcoms remained does not necessarily mean that viewers always prefer sitcoms (choice **k**); the cancelled shows might have been of very poor quality.

16. a. Choice **a** is very clearly the most accurate and thorough summary of the passage. Choices **b**, **c**, and **e** are too narrow to qualify as the main idea. The information in choice **d** is not in the passage.

17. h. Paragraph 4 states that throughout Moscow's past this square has been a commercial district. Although preservation of the past is involved, there is no support for choices **f**, **g**, or **j**. The square has never been a museum (choice **k**).

18. d. This choice is supported by the first and second sentences of paragraph 2. Choice **a** is not addressed in paragraph 2 (and is shown to be incorrect by the extensive discussion of archeological digs and historical findings). Choices **b** and **c** are not supported by the passage. Choice **e** is incorrect because Moscow, not the country of Russia, was founded in 1147.

19. h. This is the best choice because the writer speaks approvingly of both progress and preservation. Choices **f**, **g**, and **j** are refuted by the passage. There is no evidence to support choice **k**.

20. a. Choice **a** is supported by the finds described in paragraph 5. Choices **b** and **e** might be true, but the passage does not contain this information. Although the passage mentions several transportation-related items uncovered by archaeologists, there is no indication they were searching for them primarily; therefore, choice **d** is incorrect. The passage refutes choice **e**.

21. a. The main point of the passage is to explain why the arch was built and to describe the actual construction of the monument. The other choices are too vague or too narrow to be the main idea.

22. g. The second paragraph states that the Jefferson National Expansion Memorial Association held the contest to choose a design for the monument. None of the other choices are supported by the passage.

23. e. The phrase "a monument even taller than the Great Pyramid in Egypt" is evidence that the author wanted readers to understand how tall the arch is. There is no evidence in the article to support choice **a**, **c**, or **d**. Choice **b** can be ruled out because the article says, "the arch is *at least* as majestic as the Great Pyramid," not *more* majestic.

24. h. This is the best choice because the author consistently refers to the arch as "magnificent," "amazing," and "a masterpiece." The passage does not support any of the other choices.

25. a. The passage is about Thomas finally being able to write a song after his growling stomach provides him with a rhythm. Choices **b** and **d** are about details in the passages. Choices **c** and **e** contradict information in the passage.

26. g. We can conclude that Thomas is poor because he does not have any food; his refrigerator is empty. The passage suggests that Thomas takes care of himself—he attempts to feed himself when he is hungry—and there is no evidence that he doesn't take care of himself, so choice **f** is incorrect. We do not know whether Thomas had always wanted to be in a band (choice **h**). Thomas is waiting for inspiration, but there is no indication that he is waiting for someone to help him, so choice **j** is incorrect. He watches television in this passage, but we have no way of judging whether this is "too much" television or not, so choice **k** is also incorrect.

27. e. Even if you don't know that "the blues" are typically songs about hard times, the fact that Thomas used "his growling stomach" to "provide the rhythm" tells us that the song is about the hard times he has experienced on the reservation. We learn that this is not the first time he has been hungry and found his refrigerator empty; opening and closing the refrigerator was "a ceremony that he had practiced since his youth." The passage does not include any references to good times Thomas has had on the reservation, so choice **a** is incorrect. There is no information in the passage about how he and his friends started the band, so choice **b** is also incorrect. The passage does mention fry bread (choice **c**), but then the title of the song would logically have some reference to fry bread. Choice **d** is incorrect for the same reason.

28. f. Thomas refers to this opening and closing of the refrigerator as a "ceremony," and he was "expecting an immaculate conception of a jar of pickles"—a magical appearance of food, so **f** is the best answer. He is unlikely to feel disbelief that there is no food (choice **g**) because he has always had an empty refrigerator (he's performed this "ceremony" since his youth). There is no evidence that Thomas is angry (choice **h**) or that he likes the noise of the door (choice **j**), if the door indeed makes any noise. Thomas may be bored (choice **k**), but the passage doesn't tell us that.

29. d. It is ironic that in a place where there are so many ways to describe one food (indicating that this food is a central part of the culture), Thomas is hungry. The passage does not mention the language of the reservation, so choice **a** is incorrect. The sentence does not show any measure of how hungry Thomas was, so choice **b** is incorrect. The sentence does not describe fry bread or make it sound in any way appealing, so choice **c** is also incorrect. The passage tells us that it was Thomas's hunger, not the number of ways to say fry bread, that provided his inspiration, so choice **e** is incorrect.

30. c. The third paragraph states that researchers decided to improve the battery and "allow for the development of portable appliances," and the closing paragraph states that today "most homes contain about 18 different devices that use batteries." Choice **a** is a detail mentioned in the passage, not a main idea. Choices **b** and **e** are not supported by the information in the passage. Choice **d** is not mentioned.

31. f. This detail is found in the second paragraph: "only enough energy to power a dimly lit flashlight for a very short time." Choice **g** is wrong because the earlier batteries had little power. Choice **h** may be true for all types of batteries, not just the early one. All batteries have cathodes and anodes, which rules out choice **j**. Choice **k** is incorrect because there is no indication that early batteries were smaller.

32. e. This is an inference drawn from the information in the fourth paragraph. The topic sentence states that "All batteries work essentially the same way," and then goes on to say that batteries pass "electrons between the cathode and the anode." Choice **a** is incorrect because manganese dioxide is what sets alkaline batteries apart from other batteries. Choices **b**, **c**, and **d** are incorrect because they contain garbled information from the text.

33. h. This detail is drawn from the third paragraph. The researchers wanted to improve the battery so that convenience appliances could be made portable through the use of improved batteries. Choices **f**, **g**, **j**, and **k** are not supported by the information in the text.

34. a. This is an inference drawn from the entire passage: in order for batteries to become smaller and more powerful, research for their improvement must continue. Choices **b**, **c**, **d**, and **e** are inferences that are not supported by the information in the text.

35. e. The general tone of the passage indicates that neither person is very comfortable; especially note words such as "embarrassed silence," and "silence in the car becomes thicker." Choices **a** and **d** are wrong because the narrator and Jorge barely know each other and haven't ridden together before. Choices **b** and **c** are not covered by the passage.

36. j. Jorge is driving the narrator to school because the narrator's sister is ill and unable to do so. Therefore, it is likely that the narrator asked Jorge for this favor. Choice **j** is the best answer. The awkwardness of the conversation indicates that this is the first time this has happened, so choice **f** cannot be correct. If the narrator's sister had recovered from her illness (choice **g**), the entire scene would not have taken place. Jorge asked whether he has a silly laugh as the scene begins, not before the scene begins (choice **h**). The narrator notices the road construction (choice **k**) toward the end of the passage.

37. a. The narrator admits that Jorge has an annoying laugh but does not share this with Jorge because Jorge is doing the narrator a favor. The narrator's motivation is to avoid social awkwardness. The narrator makes it clear in the passage that he has no desire to be friends with Jorge (choice **b**), and there is no indication that he intends to speak to Jorge once they reach the school (choice **d**). His failure to tell Jorge the truth is a kindness rather than a punishment (choice **e**), and there is no support for the notion that he is a compulsive liar (choice **c**); in fact, he agonizes over the difference between his words and his thoughts. Choice **a** is the correct answer.

38. g. The excerpt from "Discomfort and Joy" suggests discomfort during the conversation in the car. The tension is certainly not imagined (choice **k**), although the narrator apparently feels it more strongly than Jorge does. It's true that the tension may build, as in choices **f** and **j**, or continue, as in choice **h**. However, the word *joy* in the title indicates a happier ending, so the best answer is choice **g**.

39. e. The passage specifies new research as the source of information on animal intelligence. Choices **b**, **c**, and **d** are too narrowly focused to explain the passage. Choice **a** contradicts statements in the passage.

40. g. The crow is using the stick as a tool to assist it in getting food. Choices **f**, **j**, and **k** are not signs of intelligence. Communication (choice **h**) is a sign of intelligence but is not described as involving tools.

41. e. Choice **e** is the only one mentioned in the passage as a sign of animal intelligence. Choices **a** and **d** do not indicate animal intelligence. Choices **b** and **c** might be true, but are not mentioned in the passage.

42. h. The questions in this paragraph ask the reader to consider the use of animals in our world and questions whether knowing that they have more intelligence than previously thought might make a difference in human treatment of them.

43. d. This choice is the only one that shows animals using human language. The other choices refer to sounds, not language.

44. k. The complexity of what the chimpanzee is doing to get his food and the many thinking activities he must accomplish in order to realize his goal of getting the termites—learning a new skill, selecting and shaping a tool, remembering stored knowledge, using the correct knowledge in order to take proper action for the situation—shows high intelligence.

45. b. Note that the question asks for the one answer that is NOT a sign of animal intelligence. Cuing does not demonstrate animal intelligence because the animal learns to do or not do certain things by following outside signals.

46. b. 23. Translate into an equation and solve for x: $x = 3 + (0.20)(100) = 23$.

47. f. 3.375. If $\frac{1}{16} = \frac{x}{54}$, what is x? Cross multiplying, you get $16x = 54$; then, dividing by 16, you get 3.375.

48. e. 48. Pam $= (\frac{1}{3})($Jackie$) = (\frac{1}{3})(18) = 6$, Karl $= (4)($Pam$) = (4)(6) = 24$, Pam + Karl + Jackie $= 6 + 24 + 18 = 48$.

49. j. $\frac{3}{5}$. There are 3 sections out of a total of 5 that show a number 1. That makes the probability $\frac{3}{5}$.

50. d. $p - m > n - m$. The following counterexample disproves choice **d**: $m = 6$, $n = 10$, $p = 8$, $p - m > n - m$ results in $8 - 6 > 10 - 6$ which is a false statement since 2 is not greater than 4.

51. g. $1,008. The area of Bernice's living room is $21 \times 12 = 252$ square feet. Divide by 3 to get 84 square yards. Multiply this by $12 per square yard to get $1,008.

52. c. 30 cubic feet. $V = l \times w \times h = 2 \times 3 \times 5 = 30$ cubic feet.

53. h. 17. Solve the equation: $(16)[(\frac{1}{2})(x)] = 136$, $(\frac{1}{2})(x) = 8.5$, $x = 17$.

54. d. 40 pounds. Since there are 5 parts white flour for every 1 part whole wheat flour, that means that 5 out of every 6 parts are white flour. Therefore, find $\frac{5}{6}$ of 48 pounds to see how many pounds are white flour: $\frac{5}{6} \times 48 = 40$ pounds.

55. k. 18, 17. In this series, the pattern changes from increasing by 3 to decreasing by 1. After 15, it is time to increase by 3 to 18, and then decrease by 1 to 17.

56. a. .75 days. Set up a proportion that uses ratios comparing days to supplies used: $\frac{\text{days}}{\text{amount of supplies used}} = \frac{3}{\left(\frac{2}{5}\right)} = \frac{\text{total days}}{1}$. (Use 1 because you want to see how many days it takes to use 1 full amount of supplies.) Solving this proportion, you see that they had enough food for 7.5 days, so they have 4.5 days left. Three of those days are needed for the return trip, so 1.5 remain. Divide that by 2 because half will be out and half will be back.

57. j. $55. ($80)(7) + ($20)(5) = $660 for the entire year. Divided by 12, this averages to $55 per month.

58. a. 3.
Use the Pythagorean theorem:
$a^2 + b^2 = c^2$
$1^2 + b^2 = (\sqrt{10})^2$
$b^2 = 9$ and $b = 3$

59. f. 8. Use volume to see how many smaller cubes make up the larger cube: Volume = length × width × height $= 4^3 = 64$ cubes in total. The front face and back face each have 16 cubes that will have paint on them. The top and bottom faces each have 8 cubes that are not included in the front or back face. The left and right faces each have 4 cubes that are not included in either the front/back or top/bottom faces. So the total number of cubes with paint on them will be $16(2) + 8(2) + 4(2) = 56$ cubes with paint on them. $64 - 56 = 8$ cubes will not have any paint on them.

60. b. 61. When Justina Hartley's shift in the booth began, there were $(8\frac{1}{4} \text{ pieces})(8 \text{ pies}) = \frac{33}{4} \times 8 = 66$ pieces. $66 - 5 = 61$ pieces sold.

61. h. 21 feet. Set up a proportion using ratios that compare inches to feet: $\frac{\text{inches on plan}}{\text{actual feet}} = \frac{1}{2} = \frac{1.75}{\text{actual feet}}$. Solving this gets $(12)(1.75) = 21$ feet.

62. **d.** **−28.**
$$\tfrac{3}{4}x + 7 + (-7) = -14 + (-7)$$
$$\tfrac{3}{4}x = -21$$
$$x = -21(\tfrac{4}{3}) = -28$$

63. **k.** **200.** $(-5)^2 \times (2)^3 = 25 \times 8 = 200$.

64. **a.** **145°.** Straight angles equal 180 degrees, so $180 - 35 = 145$.

65. **h.** **The area of (A) is equal to the area of (C).**
Area (A) = $2 \times 2 = 4$
Area (B) = $1 \times 3 = 3$
Area (C) = $4 \times 1 = 4$

66. **c.** **0.0025 inches.** Dividing the thickness of the book by the number of pages will show how thick each page is: $600 \div 1.5 = 0.0025$ inches.

67. **j.** **40π feet.** The longest circular path the horse can walk is the circumference around the post. $C = 2\pi r = 2\pi(20) = 40\pi$.

68. **c.** **12 minutes.** Calculating 36 psi times 20 seconds for every 1 psi will show how long it will take Sally's tire to completely deflate: $36 \times 20 = 720$ seconds. Divided by 60, this yields 12 minutes.

69. **h.** **23.** When $b = 23$, then $\sqrt{144-b} = \sqrt{144-23} = \sqrt{121} = 11$.

70. **b.** **$2(2.25) + 3 + 1.25$.** Hot dogs cost $2.25 each, a jumbo size order of french fries costs $3, and an 8-oz cup of soda costs $1.25.

71. **f.** **$24.00.** Four hamburgers will cost $10 + $3.50 = $13.50; two jumbo size fries will cost $2 \times $3 = $6; and three 12-oz sodas will cost $3 \times $1.50 = $4.50. $13.50 + $6 + $4.50 = $24.

72. **a.** **(2,−1).** Starting at the origin, the x-coordinate comes first, representing the horizontal movement, and the y-coordinate comes second, representing the vertical movement.

73. **f.** **$x^2 + 3x - 28$.** Use the acronym FOIL, which represents Firsts, Outers, Inners, Lasts, $(x + 7)(x - 4) = x^2 + (-4x) + 7x - 28 = x^2 + 3x - 28$.

74. **d.** **gallons.** Feet measure linear distances, not volume; ounces measure weight; cups measure volume, but are too small to make sense for measuring a pool, which is also true for milliliters.

75. **g.** **$283.** Kyra's employer will give her ($60)($0.10) = $6 for January; ($130)($0.10) = $13 for March, but her employer will only contribute $10 because of the maximum; and ($70)($0.10) = $7 for April. Kyra's total from her employer will be $6 + $10 + $7 = $23. Kyra's total from her own savings will be $60 + $130 + $70 = $260 and the two savings sum to $283.

76. **b.** **6 gallons.** First, make a ratio representing the rate of rain in terms of (# gallons): (# hours). It will be (6 gallons):(24 hours) and set that equal to (x gallons):(8 hours). $\frac{6}{24} = \frac{x}{8}$, so 2 gallons would have been collected in the barrel after 8 hours of rain. Adding to the original 4 gallons shows that 6 gallons were in the tank after the thunderstorm.

77. **k.** **18.** Circumference $= 2\pi r = 36\pi$, so $r = 18$.

78. **b.** **4 cm.** Find the missing side on box B.
Volume of Box A = $4 \text{ cm} \times 4 \text{ cm} \times 3 \text{ cm} = 48 \text{ cm}^3$
Volume of Box B = $2 \text{ cm} \times 6 \text{ cm} \times x \text{ cm} = 12x \text{ cm}^3$, so $x = 4$ cm

79. **f.** **Fewer DVDs were sold on the last three days than on the first two days.** Wed + Thurs + Fri = 150 + 75 + 100 = 325. Mon + Tues = 125 + 225 = 350. There were fewer DVDs sold on the last 3 days than on Monday and Tuesday.

80. **d.** **If it does not rain, then I do not take an umbrella.** The inverse of the statement, "If a, then b," is, "If not a, then not b."

81. **h.** **7.** $12a + \frac{b^2}{4} = 93$, $12a + \frac{6^2}{4} = 12a + \frac{36}{4} = 12a + 9 = 93$, $12a = 93 - 9 = 84$, so $a = 7$.

82. e. 43,000. $3{,}400 + 8{,}500 + 7{,}700 + 5{,}600 + 7{,}100 + 11{,}300 = 43{,}600$.

83. f. $32x^{11}y^5$. First, when raising a power to a power as in $(2x^3y)^3$, the exponent gets distributed to all terms in the parenthesis and multiplies with those exponents. Therefore, $(2x^3y)^3 = (2^3)(x^3)^3(y)^3 = 8x^9y^3$. When multiplying terms with like bases, add exponents of like bases and multiply coefficients: $(8x^9y^3)(4x^2y^2) = 32x^{11}y^5$.

84. c. $\frac{8}{5}$. Take the square root of the numerator and denominator each separately: $\frac{\sqrt{64}}{\sqrt{25}} = \frac{8}{5}$.

85. k. $\frac{3}{11}$. The number of blue parts will be 450: $1650 - (430 + 425 + 345) = 450$. Then using $\frac{\text{number of desired events}}{\text{total number of events}}$ the probability of getting a blue piece is $\frac{450}{1{,}650} = \frac{45}{165} = \frac{9}{33} = \frac{3}{11}$.

86. d. 7 hours, 12 minutes. Set up a proportion: $\frac{50\text{-gallon}}{5\text{ hours}} = \frac{72\text{-gallon}}{x\text{ hours}}$. Solving for x gives $x = 7.2$ hours. 0.2 of 60 minutes is $(0.2)(60) = 12$ minutes.

87. h. 18. 3 apples + 6 oranges = 9 pieces of fruit, so there are 18 cherries.

88. a. $2,336.67. Use multiplication to find the percentage of the second week's sales, which were $73,021: $(3.2\%)(\$73{,}021) = (0.032)(73{,}021) = \$2{,}336.67$.

89. g. 61.5°. The median is the middle term when the terms are ordered from least to greatest. When there is an even number of terms, the center two terms are averaged: 60, 60, 61, 62, 64, 65 is the proper ordering, so take the average of 61 and 62 which is 61.5.

90. d. 807. First change the mile into inches. 1 mile = (5,280 feet)(12 inches) = 63,360 inches. Using circumference we see that for each turn, the wheel travels 25π, or 78.5 inches. Lastly, $\frac{63{,}360}{78.5} = 807.1$ turns.

91. k. 252 in.² (11 in.)(8 in.)(2 sides) + (11 in.)(2 in.)(2 sides) + (2 in.)(8 in.) (2 sides) = 252 in².

92. c. $9.79. $\frac{\$5.34}{6} = \$0.89$ per can. $11 \times \$0.89 = \$9.79$.

93. f. 1. Prime numbers must be greater than 1 and must only have 2 factors: 1 and themselves. 2 is the only even prime number.

94. b. $675 Canadian. $(\$500)(1.35) = \$675$ Canadian.

95. h. 30. Begin by listing the multiples of 15 until you get to one that 6 is also a multiple of: 15, 30, 45, 60 . . . 30 is the first number that 6 is also a multiple of.

6 ▶ PRACTICE TEST 4

The *SHSAT Power Practice* tests will help you prepare for the high-stakes exams given to students applying for New York City's specialized high schools. Each practice test consists of sample questions like those you will find on the official SHSAT.

The 45-question verbal section and 50-question math section were developed by education experts. These tests will show you how much you know and what kinds of problems you still need to study. Mastering these practice tests will allow you to reach your highest potential on the real SHSAT.

PART I VERBAL

Scrambled Paragraphs

Paragraph 1
q r s t u
q r s t u
q r s t u
q r s t u
q r s t u

Paragraph 2
q r s t u
q r s t u
q r s t u
q r s t u
q r s t u

Paragraph 3
q r s t u
q r s t u
q r s t u
q r s t u
q r s t u

Paragraph 4
q r s t u
q r s t u
q r s t u
q r s t u
q r s t u

Paragraph 5
q r s t u
q r s t u
q r s t u
q r s t u
q r s t u

Logical Reasoning

6. a b c d e
7. f g h j k
8. a b c d e
9. f g h j k
10. a b c d e
11. f g h j k
12. a b c d e
13. f g h j k
14. a b c d e
15. f g h j k

Reading

16. a b c d e
17. f g h j k
18. a b c d e
19. f g h j k
20. a b c d e
21. f g h j k
22. a b c d e
23. f g h j k
24. a b c d e
25. f g h j k
26. a b c d e
27. f g h j k
28. a b c d e
29. f g h j k
30. a b c d e

31. f g h j k
32. a b c d e
33. f g h j k
34. a b c d e
35. f g h j k
36. a b c d e
37. f g h j k
38. a b c d e
39. f g h j k
40. a b c d e
41. f g h j k
42. a b c d e
43. f g h j k
44. a b c d e
45. f g h j k

PART II MATHEMATICS

46. a b c d e
47. f g h j k
48. a b c d e
49. f g h j k
50. a b c d e
51. f g h j k
52. a b c d e
53. f g h j k
54. a b c d e
55. f g h j k
56. a b c d e
57. f g h j k
58. a b c d e
59. f g h j k
60. a b c d e
61. f g h j k
62. a b c d e

63. f g h j k
64. a b c d e
65. f g h j k
66. a b c d e
67. f g h j k
68. a b c d e
69. f g h j k
70. a b c d e
71. f g h j k
72. a b c d e
73. f g h j k
74. a b c d e
75. f g h j k
76. a b c d e
77. f g h j k
78. a b c d e
79. f g h j k

80. a b c d e
81. f g h j k
82. a b c d e
83. f g h j k
84. a b c d e
85. f g h j k
86. a b c d e
87. f g h j k
88. a b c d e
89. f g h j k
90. a b c d e
91. f g h j k
92. a b c d e
93. f g h j k
94. a b c d e
95. f g h j k

Part 1—Verbal

The Verbal Test includes 45 questions in these three sections:

- Scrambled Paragraphs, 5 paragraphs (each counts double)
- Logical Reasoning, 10 questions, numbered 6–15
- Reading, 30 questions, numbered 16–45

Scrambled Paragraphs

This section tests your ability to organize a paragraph well. There are five paragraphs, presented in scrambled order. Your job is to put them in the best order to make a clear, coherent paragraph. Each correct answer counts double; these five paragraphs are worth 10 points out of the 50-point verbal test.

The first sentence in each paragraph is given. The remaining five sentences are listed in random order. Read each group of sentences carefully, and then decide on the best arrangement for them. Use the blanks at the left of each sentence to number these sentences from 1 to 5, showing the order they should be in.

Paragraph 1

Howard found his science classroom both spooky and fascinating.

_____ **Q.** The classroom was in the basement of the General Sciences Building.

_____ **R.** The cases contained exhibits whose titles read "Reptiles of the Desert Southwest" and "Birds of the Central United States."

_____ **S.** Below the stands, a typewritten card, yellow with age, bore the name of each bird's genus and species.

_____ **T.** The dusty specimens inside the bird cases were perched on little stands, their tiny claws gripping the smooth wood nervously.

_____ **U.** The shadowy corridor that led back to it was lined with glass cases.

Paragraph 2

Batman was the brainchild of comic book artist Bob Kane.

_____ **Q.** The comic book company asked Kane for a new character, just as powerful as Superman, to appeal to its readers.

_____ **R.** DC contacted Kane because their Superman hero was already a phenomenal success.

_____ **S.** Kane's ideas for Batman reportedly came from Leonardo da Vinci's famous sketch of a man flying with batlike wings, and was also said to be influenced by heroes in the *Shadow* and *Zorro* series who wore masks.

_____ **T.** Kane responded to DC's request with his ideas about a character called Batman.

_____ **U.** Kane was just 22 years old when he was asked to create a new superhero for DC Comics.

Paragraph 3

Bob Kane didn't give his superhero, Batman, any supernatural powers.

_____ **Q.** Thus, Kane gave Batman physical strength and unique weapons, but made him just as human as the rest of us.

_____ **R.** Wayne also used his wealth to develop high-tech crime-fighting tools and weapons, like his famous Batmobile.

_____ **S.** He vowed to avenge their deaths and dedicated himself to bringing all criminals to justice.

_____ **T.** In Kane's story, Batman is really Bruce Wayne, a human millionaire who witnessed the murder of his parents as a child.

_____ **U.** To fulfill his vow, Bruce Wayne devoted his life to training his body and mind to fight crime as Batman.

Paragraph 4

He drove home by his usual route, a road he had taken a thousand times.

_____ **Q.** He had taken his daughter Abigail there and taught her how to skate when she was small.

_____ **R.** Still, he did not know why, as he passed the park, he should so suddenly and vividly picture the small pond that lay at the center of it.

_____ **S.** She was wearing her new skates, going much faster than she should have been.

_____ **T.** Now there came into his mind an image of Abigail gliding toward him.

_____ **U.** In winter this pond was frozen over.

Paragraph 5

Whether you are evaluating an entire debate team or just one speaker, there are several things to keep in mind.

_____ **Q.** Also pay attention to the speaker's use of wit, repetition, and other rhetorical devices that can grab and hold the audience's attention.

_____ **R.** Delivery refers to the speaker's style and clarity.

_____ **S.** Content refers to the logic of the arguments, the effectiveness of any factual support, and the thoroughness of the rebuttals.

_____ **T.** As you listen to the debate, you may find it helpful to consider the two attributes of delivery and content.

_____ **U.** First of all, you must be careful not to let your own opinions on the issues interfere with your assessment of the debaters.

Logical Reasoning

The questions in this section test your ability to reason well, that is, to figure out what the facts you know can or can't possibly mean. Read the statements carefully, then choose the best answer based *only* on the information given. Note carefully the words used in each question. For example, one thing can be lar*ger* than another without being the lar*gest* in the group. In answering some of these questions, it may be useful to draw a rough diagram or make a list that gives real world values to the information.

6. Phil lives 14 hours from New York City. He drove eight hours on Monday, and six hours on Tuesday.

   Which of the following can we conclude is true?

   a. Phil drove 14 hours over two days.
   b. Phil arrived in New York City on Tuesday.
   c. Phil drives fast.
   d. Phil spent Monday night at a hotel.
   e. Phil doesn't visit New York City often.

7. The Greenwood family has four dogs.
   ■ Dottie is bigger than King and smaller than Sugar.
   ■ Ralph is smaller than Sugar and bigger than Dottie.

   Which is the largest of the four dogs?

   f. Sugar
   g. Dottie
   h. King
   j. Ralph
   k. It cannot be determined from the information given

8. Francesca and Mario shared an art space for a photography exhibition in New Orleans. Mario sold four of his works and Francesca sold six.

Based on the information above, which statement or statements must be true?

   a. Mario and Francesca are the best photographers in New Orleans.
   b. Mario sold fewer works than Francesca.
   c. Francesca is a better photographer than Mario.
   d. Mario bought one of Francesca's works.
   e. Francesca taught Mario how to take photographs.

9. The Smiths own a small house with a large garden, two cars, and a small garage. Mr. and Mrs. Smith drive a blue Jeep Cherokee, which they park in their garage every night. Their son, Josh, has a black Nissan, which he keeps in the driveway.

Based on this information, we can conclude that:

   f. The Smiths own more than two cars.
   g. The Smiths have a one-car garage.
   h. Josh uses his car more often than his parents use theirs.
   j. The Smiths like their car more than Josh likes his.
   k. It cannot be determined from the information given.

Read the following and answer Questions 10 and 11.

The following code has the following rules:

   1. Each letter in the code represents the same word in all three sentences.
   2. Each word is represented by only one letter.
   3. The position of a letter in any of the sentences is never the same as that of the word it represents.

A E I O U means:
   "Twenty people attended the meeting."

U Z O A E means:
   "Thirty people attended the meeting."

O I E X A means:
   "Twenty students attended the meeting."

10. Which letter represents the word "Thirty"?
   a. Z
   b. X
   c. I
   d. A
   e. It cannot be determined from the information given.

11. Which word is represented by the letter **U**?
   f. meeting
   g. students
   h. attended
   j. people
   k. It cannot be determined from the information given.

12. Meghan: While I was at the pet store, every rabbit that bit me was a gray rabbit, and every gray rabbit I touched bit me.

Based on Meghan's statement, which of the following statements can be correctly inferred?

   a. The only gray animals that Meghan saw at the pet store were rabbits.
   b. There were no white rabbits at the pet store.
   c. While at the pet store, no white rabbits bit Meghan.
   d. Every rabbit that Meghan touched at the pet store bit her.
   e. The pet store had only rabbits and no other animals.

13. To enter a creative writing program, all students must submit a 50-page portfolio of original work. Veronika is in a creative writing program.

Which of the following can we conclude is true?

f. Veronika submitted a 30-page portfolio of original work.

g. Veronika is an excellent student.

h. Veronika does not enjoy her program.

j. Veronika is a better writer than most students in her program.

k. Veronika submitted a 50-page portfolio of original work.

14. By the time Roma approached his building, he remembered that he had forgotten his apartment keys at the office. Through the windows of his apartment, he saw that the lights were on in the kitchen, but when he rang the doorbell, no one answered.

Which of the following must be true?

a. Roma could not get into his apartment.

b. Somebody was in the kitchen.

c. Roma has a spare set of keys at his office.

d. Whoever was in Roma's apartment did not want to open the door for Roma.

e. It cannot be determined from the choices offered.

15. When Anya does not have a full-time job, she spends her time playing music all over New York City. When she works full-time, she does not perform at all. Last February, Anya performed nearly every day.

Which of the following can we assume is true?

f. Anya quit her job at the end of January.

g. Anya would prefer not to have a full-time job.

h. No one performed as much as Anya last February.

j. Anya is one of the best musicians in New York City.

k. Last February, Anya did not have a full-time job.

Reading Comprehension

This section tests your reading comprehension—your ability to understand what you read. Read each passage carefully and answer the questions that follow it. If necessary, you can reread the passage to be certain of your answers. Remember that your answers must be based only on information that is actually in the passage.

Read the following passage, then answer Questions 16 through 20.

The coast of the state of Maine is one of the most irregular in the world. A straight line running from the southernmost coastal city to the northernmost coastal city would measure about 225 miles. If you followed the coastline between these points, you would travel more than ten times as far. This irregularity is the result of what is called a drowned coastline. The term comes from the glacial activity of the ice age. At that time, the whole area that is now Maine was part of a mountain range that towered above the sea. As the glacier descended, however, it expended enormous force on those mountains, and they sank into the sea.

As the mountains sank, ocean water charged over the lowest parts of the remaining land, forming a series of twisting inlets and lagoons of contorted grottos and nooks. The highest parts of the former mountain range, nearest the shore, remained as islands. Mt. Desert Island is one of the most famous of all the islands left behind by the glacier. Marine fossils found here were 225 feet above sea level, indicating the level of the shoreline prior to the glacier.

The 2,500-mile-long rocky and jagged coastline of Maine keeps watch over nearly two thousand islands. Many of these islands are tiny and uninhabited, but many are home to thriving communities. Mt. Desert Island is one of the largest, most beautiful of the Maine coast islands. Measuring 16 miles by 12 miles, Mt. Desert was essentially formed as two distinct islands. It is split almost in half by Somes Sound, a deep and narrow stretch of water, seven miles long.

For years, Mt. Desert Island, particularly its major settlement, Bar Harbor, afforded summer homes for the wealthy. Recently, though, Bar Harbor has become a burgeoning arts community as well. But, the best part of the island is the unspoiled forest land known as Acadia National Park. Because the island sits on the boundary line between the temperate and sub-Arctic zones, the island supports the flora and fauna of both zones as well as beach, inland, and alpine plants. It also lies in a major bird migration lane and is a resting spot for many birds. The establishment of Acadia National Park in 1916 means that this natural reserve will be perpetually available to all people, not just the wealthy. Visitors to Acadia may receive nature instruction from the park naturalists as well as enjoy camping, hiking,

cycling, and boating. Or they may choose to spend time at the archeological museum, learning about the Stone Age inhabitants of the island.

The best view on Mt. Desert Island is from the top of Cadillac Mountain. This mountain rises 1,532 feet, making it the highest mountain on the Atlantic seaboard. From the summit, you can gaze back toward the mainland or out over the Atlantic Ocean and contemplate the beauty created by a retreating glacier.

16. Which of the following best tells what this passage is about?
   a. The islands along the coastline of Maine enchant residents and visitors alike. Visitors can receive nature instruction from Arcadia park naturalists.
   b. Only wealthy summer vacationers who have homes in Bar Harbor ever visit Maine's beautiful Mt. Desert Island.
   c. The tops of a former mountain range became islands off the Maine coast.
   d. Marine fossils found on Mt. Desert Island were 225 feet above sea level; that was the level of the shoreline prior to the glacier.
   e. Natural forces formed Maine coastal islands; today's Mt. Desert Island is a beautiful place, popular with residents and tourists.

17. According to the passage, the large number of small islands along the coast of Maine are the result of
   f. glaciers forcing a mountain range into the sea.
   g. Maine's location between the temperate and sub-Arctic zones.
   h. the irregularity of the Maine coast.
   j. the need for summer communities for wealthy tourists and artists.
   k. the incredibly long coastline.

18. The content of the fourth paragraph indicates that the writer believes that
    a. the continued existence of national parks is threatened by budget cuts.
    b. the best way to preserve the environment on Mt. Desert Island is to limit the number of visitors.
    c. national parks allow large numbers of people to visit and learn about interesting wilderness areas.
    d. Mt. Desert Island is the most interesting tourist attraction in Maine.
    e. Mt. Desert Island should be closed to tourists in order to preserve its delicate ecological system.

19. According to the selection, the coast of Maine is
    f. 2,500 miles long.
    g. 3,500 miles long.
    h. 225 miles long.
    j. 235 miles long.
    k. 522 miles long

20. Which of the following statements best expresses the main idea of the fourth paragraph of the selection?
    a. The wealthy residents of Mt. Desert Island selfishly kept it to themselves.
    b. Acadia National Park is one of the smallest of the national parks.
    c. On Mt. Desert Island, there is great tension between the year-round residents and the summer tourists.
    d. Due to its location and environment, Mt. Desert Island supports an incredibly diverse animal and plant life.
    e. Many migrating birds rest on the island.

Read the following passage, then answer Questions 21 through 25.

For eight weeks last summer, scientists from several universities and government weather laboratories conducted the Severe Thunderstorm Electrification and Precipitation Study (STEPS). They set up shop in Goodland, Kansas, home to some of the most violent thunderstorms in the United States. In this area, moist air from the Gulf of Mexico meets hot, dry air from the southwest, resulting in storms so huge that they can last for days as they move east across the country.

The scientists working in Goodland were trying to learn as much as they could about these chaotic storm systems. They already knew that the combination of strong winds, large amounts of moisture, and significant differences in temperatures could produce thunderheads with vortices of circulating air, known as super cells. They also knew that supercells could produce violent weather, including tornadoes. What the scientists did not know, however, was why some supercells produce heavy rain, lightning, and tornadoes, while others produce large hail that can ruin a wheat crop or damage a roof. The scientists suspected that the physical properties of storm clouds would give the reason for such differences.

To study the storm clouds, the researchers in Goodland met each morning and studied weather data, hoping for bad weather. Their cars, complete with roof-mounted weather instruments, were ready. Whenever a storm approached, a crew of scientists drove toward the storm front and begin collecting data on winds, temperature, barometric pressure, and humidity. During the storm, they also calculated the time and place of every lightning strike.

At the same time that the scientists on the ground were gathering their data, researchers in the air were operating a special radar station that would help them measure the shape and size of the water particles inside the clouds. These scientists rode in a single-engine plane specifically designed to fly through severe weather systems. While the pilot guided the plane through heavy wind, lightning, hail, and ice, the scientists on board gathered their storm data.

Researchers admit that it will take years for them to understand all the data they have collected. Their hope is that by comparing the conditions on the ground with the conditions in the air, it may be possible to learn much more about how lightning is generated, why some storms dump hail the size of baseballs, and why others produce devastating floods and tornadoes.

21. The answer to which of the following questions would best tell what this passage is about?
    a. How did scientists apply what they knew about storms to their research?
    b. What made scientists choose Goodland, Kansas, as their research base?
    c. How did the data collected on land compare to that collected in the air?
    d. What is the difference between a thunder-storm and a lightning storm?
    e. How did scientists attempt to learn more about severe storm systems?

22. Which of the following actions did the scientists carry out first?
    f. They gathered storm data.
    g. They formed a hypothesis.
    h. They drove toward the storm front.
    j. They waited for bad weather.
    k. They measured water particles.

23. Which of the following is suggested about the research data collected by the scientists in Kansas?
    a. The data has explained why some storms dump hail and others do not.
    b. The data was so great that it will take time to sort out properly.
    c. The data has provided clear answers to the scientists' questions.
    d. The data will trigger other research projects at the STEPS laboratories.
    e. The data contradicted what most researchers suspected about supercells.

24. The passage states that large, chaotic storm systems can be caused
    f. by tornadoes.
    g. by large hail and ice.
    h. when water particles make clouds grow too large.
    j. when barometric pressure rises and humidity lowers.
    k. when moist air and hot, dry air clash.

25. Some of the scientists rode in a special single-engine plane mainly because they
    a. wanted to view the storms from another angle.
    b. needed to communicate with the researchers on the ground.
    c. hoped to gather data about the composition of storm clouds.
    d. were the only ones who could operate the special radar station.
    e. hoped to be able to see exactly how hail is formed.

Read the following passage, then answer Questions 26 through 30.

[In this excerpt, Jane Eyre decides to leave Lowood, the boarding school where she has lived for eight years.]

My world had for some years been in Lowood: my experience had been of its rules and systems; now I remembered that the real world was wide, and that a varied field of hopes and fears, of sensations and excitements, awaited those who had courage to go forth into its expanse, to seek real knowledge of life amidst its perils.

I went to my window, opened it, and looked out. There were the two wings of the building; there was the garden; there were the skirts of Lowood; there was the hilly horizon. My eye passed all other objects to rest on those most remote, the blue peaks: it was those I longed to surmount; all within their boundary of rock and heath seemed a prison-ground, exile limits. I traced the white road winding round the base of one mountain, and vanishing in a gorge between two: how I longed to follow it further! I recalled the time when I had traveled that very road in a coach; I remembered descending that hill at twilight: an age seemed to have elapsed since the day which brought me first to Lowood, and I had never quitted it since. My vacations had all been spent at school: Mrs. Reed had never sent for me to Gateshead; neither she nor any of her family had ever been to visit me. I had had no communication by letter or message with the outer world: school-rules, school-duties, school-habits and notions, and voices, and faces, and phrases, and costumes, and preferences, and antipathies: such was what I knew of existence.

And now I felt that it was not enough: I tired of the routine of eight years in one

afternoon. I desired liberty; for liberty I gasped; for liberty I uttered a prayer; it seemed scattered on the wind then faintly blowing. I abandoned it and framed a humbler supplication; for change, stimulus: that petition, too, seemed swept off into vague space: "Then," I cried, half desperate, "grant me at least a new servitude!"
—Charlotte Bronte, from *Jane Eyre* (1847)

26. Which of the following best tells what this passage is about?
    a. Jane finds the idea of the world beyond Lowood frightening and wonders if she can ever leave.
    b. Looking at the blue mountain peaks, Jane longs to climb them.
    c. Comparing the world of Lowood with what she knows of the outside world, Jane is eager to leave the boarding school.
    d. Jane remembers vacations away from school and decides she will leave.
    e. Jane considers a letter from a friend urging her to leave Lowood.

27. Which of the following best describes life at Lowood?
    f. very unconventional and modern
    g. very structured and isolated
    h. harsh and demeaning
    j. liberal and carefree
    k. urban and sophisticated

28. The narrator
    a. realizes she wants to experience the world.
    b. decides that she must get married.
    c. realizes she can never leave Lowood.
    d. decides to return to her family.
    e. determines to become a teacher.

29. Which of the following best describes Jane's nature?

    f. fearful and angry

    g. timid and solitary

    h. dependent on her friends

    j. steady and sensible

    k. curious and adventurous

30. The narrator reduces her prayer from freedom to "at least a new servitude." Judging from the passage, which of these is most likely the reason that she changes her prayer?

    a. She doesn't believe in prayer.

    b. She is not in a free country.

    c. She has been offered a position as a servant.

    d. She thinks that freedom might be an unrealistic wish.

    e. She has been treated like a slave at Lowood.

Read the following passage, then answer Questions 31 through 36.

Although the Gunnison sage grouse—a bird that was named for the area of Colorado where it is most plentiful—has been around at least as long as the northern sage grouse, it has only recently received recognition as a new species. Bird taxonomists had assumed that the Gunnison sage grouse and the northern sage grouse were the same species, even though the two are very different in size and behavior. Recent DNA testing, however, demonstrated that the birds are too distantly related to be the same species.

Even before the DNA tests, there were clues that the Gunnison was not related to the northern sage grouse. For one thing, the two species do not interbreed. In fact, when one scientist, Dr. Jessica Young, played a recording of the mating calls of the male Gunnison sage grouse, the female northern sage grouse immediately left the area. The same thing happened when Dr. Young played northern sage grouse sounds to the female Gunnison sage grouse.

The two species are also very different in appearance. The Gunnison sage grouse is smaller than the northern sage and has different coloration. For example, the Gunnison has a white band on its tail feathers that the northern sage grouse does not have. The two are also very different in behavior. Gunnison males have a different kind of strut and like to throw their elaborate neck feathers over their heads and wag their tail feathers in a visible display.

The Gunnison sage grouse will now be added to the American Ornithological Union's official catalog of bird species. Officials at the Union believe it is the first time a new species has been added since the end of the nineteenth century, and bird-watchers are anxious to make trips to Colorado's Gunnison Basin, hoping to view the new species.

Unfortunately, just as the Gunnison sage grouse is receiving recognition as a distinct species, it may be time to add it to the endangered species list. Environmental groups are worried about the bird's survival and have petitioned the U.S. Fish and Wildlife Service for an emergency listing. Experts estimate that the northern sage grouse population has declined by 90% or more—from about 1 million birds down to about 100,000. Environmentalists believe the situation of the Gunnison sage grouse is even worse.

31. Which of the following best tells what this passage is about?

    a. a suggestion that there are unknown species of sage grouse

    b. a comparison and contrast of two species of sage grouse

    c. a plea to save Colorado's endangered sage grouse

    d. a theory about sage grouse that has been proven to be true

    e. a prediction that new sage grouse species will soon be discovered

32. Based on the passage, how are the Gunnison sage grouse and the northern sage grouse alike?

    f. They have the same coloration.

    g. The two species interbreed.

    h. They are similar in size and weight.

    j. Their populations have been declining.

    k. They have the same strut and the same tail feathers.

33. The passage suggests that the Gunnison sage grouse has probably existed

    a. ever since DNA tests demonstrated it was distinct.

    b. as long as any other species of sage grouse.

    c. ever since the end of the nineteenth century.

    d. as long as there has been an endangered species list.

    e. ever since people settled the Gunnison Basin.

34. Which of the following proved that the Gunnison sage grouse and the northern sage grouse were different species?

    f. Taxonomists cited the birds' different sizes.

    g. Dr. Young's studies revealed that the birds had different tail feathers.

    h. DNA testing indicated that the birds were not closely related.

    j. The Gunnison sage grouse was put on the endangered species list.

    k. The birds' populations showed markedly different rates of decline.

35. The discussion of Dr. Young's research suggests that

    a. mating only occurs within, and not among, sage grouse species.

    b. female sage grouse do not recognize male mating calls.

    c. sage grouse mating habits explain why the species are endangered.

    d. Gunnison sage grouse have a mating call but northern sage grouse do not.

    e. DNA has little or nothing to do with species' mating habits.

36. Environmentalists want the Gunnison sage grouse added to the endangered species list because

    f. its misidentification with another species has threatened its survival.

    g. its survival rate, according to U.S. Fish and Wildlife, is only 90%.

    h. there are great numbers of bird-watchers anxious to view the species.

    j. its total population is only about 1 million birds.

    k. its total population is probably less than 100,000 birds.

Read the following passage, then answer Questions 37 through 40.

Barney and Judy lived near each other, but they were never allowed to spend time together. They were both very lonely, but were forced to live alone. After hearing about their plight, a group of lawyers decided to file a lawsuit on their behalf.

What is unusual about the lawsuit is that Barney and Judy are chimpanzees who live at a zoo. The chimps' lawyers sued their zookeepers under amendments to the Federal Animal Welfare Act, which states that zoos that confine primates must safeguard their psychological as well as their physical well-being. Judy and Barney won their lawsuit, a groundbreaking ruling on animal rights, and were allowed to visit with other chimpanzees.

Until very recently, when the law dealt with animals, it viewed them merely as someone's property. But as anyone who has ever had a beloved pet will tell you, animals are much more than property. Not only have the relationships between people and animals changed over the generations, but research has given us greater insight into the minds of animals. Most researchers now believe that animals experience not only pain, but sadness; not only hunger, but loneliness; not only self-interest, but true affection. It is becoming increasingly difficult to think of such creatures as mere property.

Not everyone approves of this new field of legal practice called animal law. Some argue that human society is defined largely by its differentiation from animal society. The more that line is blurred, the more human society will lose its denotation. Nonetheless, the movement to defend animal rights is growing. In addition, the new animal rights attorneys point out that lawyers fighting for civil rights and environmental protections were ridiculed when they first began those legal battles.

One sign that the rights of animals are being taken more seriously is that, in 1994, 44 states defined animal cruelty as a misdemeanor crime; by 1999, more than half the states had made animal cruelty a felony, a much more serious crime. Another sign is the growing number of lawyers practicing animal law. The Animal Defense Fund, based in Petaluma, California, has an annual budget of $3 million. A five-lawyer firm in Washington, DC, specializes in animal rights and recently handled a complex case involving standards for the treatment of circus elephants. Lawyers such as these have essentially created an entirely new field of law.

37. Which of the following best tells what this passage is about?
   a. scientific research indicating that animals have feelings
   b. the development of a new field of law called animal law
   c. the theory that chimpanzees need psychological care
   d. the effect animal rights activists can have on animal rights
   e. the large number of lawyers who now specialize in animal law

38. According to the passage, why did the lawyers sue the chimpanzees' zookeepers?
   f. Barney and Judy were ill and were not being fed properly.
   g. The lawyers were eager to win a groundbreaking ruling on animal rights.
   h. Barney and Judy's caregivers viewed them as mere property of the zoo.
   j. The lawyers had visited the zoo and had seen that the chimps were lonely.
   k. Barney and Judy's psychological health was at risk.

39. Which of the following most accurately states the result of Barney and Judy's lawsuit?

　a. The chimps' lawyers became leaders in the practice of animal law.

　b. Researchers found that pets experience a range of human feelings.

　c. The zoo was no longer allowed to keep Barney and Judy.

　d. Barney and Judy may now spend time with other chimpanzees.

　e. Money that was awarded in the chimps' lawsuit established an animal rights fund.

40. With which of the following statements would Barney and Judy's lawyers probably *disagree*?

　f. Household pets should be given special protection under the law.

　g. Household pets should be treated as though they have feelings.

　h. A person owns a pet in the same way that he or she owns property.

　j. Animals are social creatures and can experience loneliness if they are isolated.

　k. Animals sometimes need legal protection.

Read the following passage, then answer Questions 41 through 45.

Book clubs are a great way to meet new friends or keep in touch with old ones, while keeping up on your reading and participating in lively and intellectually stimulating discussions. If you're interested in starting a book club, you should consider the following options and recommendations.

The first thing you'll need are members. Before recruiting, think carefully about how many people you want to participate and also what the club's focus will be. For example, some book clubs focus exclusively on fiction, others read nonfiction. Some are even more specific, focusing only on a particular genre such as mysteries, science fiction, or romance. Others have a more flexible and open focus. All these possibilities can make for a great club, but it is important to decide on a focus at the outset so the guidelines will be clear to the group and prospective members.

After setting the basic parameters, recruitment can begin. Notify friends and family, advertise in the local newspaper, and hang flyers on bulletin boards in local stores, colleges, libraries, and bookstores. When enough people express interest, schedule a kickoff meeting during which decisions will be made about specific guidelines that will ensure the club runs smoothly. This meeting will need to establish where the group will meet (rotating homes or a public venue such as a library or coffee shop); how often the group will meet, and on what day of the week and at what time; how long the meetings will be; how books will be chosen and by whom; who will lead the group (if anyone); and whether refreshments will be served and, if so, who will supply them. By the end of this meeting, these guidelines should be set and a book selection and date for the first official meeting should be finalized.

Planning and running a book club is not without challenges, but when a book club is run effectively, the experience can be extremely rewarding for everyone involved.

41. Which of the following titles best tells what this passage is about?

　a. Book Clubs: A Great Way to Make New Friends

　b. Starting a Successful Book Club: A Guide

　c. Five Easy Steps to Starting a Successful Book Club

　d. Books versus Computers

　e. Reading in Groups: Sharing Knowledge, Nurturing Friendships

42. According to the passage, when starting a book club, the first thing a person should do is
  f. hang flyers in local establishments.
  g. put an ad in a local newspaper.
  h. decide on the focus and size of the club.
  j. decide when and where the group will meet.
  k. elect officers.

43. Which of the following would NOT be covered during the book club's kickoff meeting?
  a. deciding on whether refreshments will be served
  b. discussing and/or appointing a leader
  c. choosing the club's first selection
  d. finalizing a date for the first official meeting
  e. identifying what kinds of books or genre will be the club's focus

44. Which of the following is NOT something that successful book clubs should do?
  f. focus exclusively on one genre
  g. have guidelines about where and when to meet
  h. have a focus
  j. decide how to choose and who will choose book selections
  k. hang flyers on bulletin boards

45. Which of the following inferences can be drawn from the passage?
  a. Smaller groups are better for a variety of reasons.
  b. The social aspect of book clubs is more important than the intellectual.
  c. Starting your own book club is better than joining an existing one.
  d. When starting and running a book club, a casual approach is risky.
  e. You should choose a meeting place where refreshments are available.

Part 2—Math

The Math Test includes 50 questions covering content in the following areas:

- basic math
- percentages, fractions, decimals, averages
- pre-algebra
- algebra
- substitution
- factoring
- geometry
- probability
- logic
- word problems

Solve each problem and select the best answer from the choices given. It is important to keep in mind that:

- Formulas and definitions of mathematical terms and symbols are not provided.
- Diagrams other than graphs are not necessarily drawn to scale. Do not assume any relationship in a diagram unless it is specifically stated or can be figured out from the information given.
- A diagram is in one plane unless the problem specifically states that it is not.
- Graphs are drawn to scale. Unless stated otherwise, you can assume relationships according to appearance. For example, (on a graph) lines that appear to be parallel can be assumed to be parallel; likewise for concurrent lines, straight lines, collinear points, right angles, and so on.
- You will need to reduce all fractions to lowest terms.

46. Find the median of the following series of numbers:

8, 6, 8, 9, 4, 8, 12, 9

   a. 6.5
   b. 7
   c. 8
   d. 9
   e. 12

47. The equation $2y + 10 = x$ defines
   f. a line with slope = 2.
   g. a line with slope = $\frac{1}{2}$.
   h. a line with no slope.
   j. a line with slope = 5.
   k. a line with negative slope.

48. Which polynomial should be added to $4x^2 + 5x - 8$ in order to equal to 0?
   a. $4x + 5x + 8$
   b. $-4x^2 - 5x + 8$
   c. $-4x^2 - 5x - 8$
   d. $-4x^2$
   e. $4x^2 + 5x - 8$

49. Which is the sum of the areas of the rectangles shown here as a polynomial in simplest form?

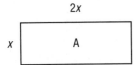
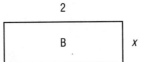
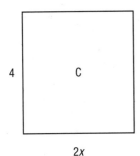
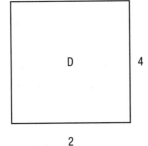

   f. $8x + 4$
   g. $x^2 + 8x$
   h. $2x^2 + 2x + 4$
   j. $2x^2 + 10x + 8$
   k. $4x^2 + 12x + 4$

50. What is the y-intercept of the line defined by $4y = 12x + 4$?
   a. (0,4)
   b. (4,0)
   c. (0,1)
   d. (1,0)
   e. (0,3)

51. Please use the following to answer Question 51:

DISTANCE TRAVELED FROM CHICAGO WITH RESPECT TO TIME	
Time (hours)	Distance from Chicago (miles)
1	60
2	120
3	180
4	240

A train moving at a constant speed leaves Chicago for Los Angeles at time $t = 0$. If Los Angeles is 2,000 miles from Chicago, which of the following equations describes the distance from Los Angeles at any time t?
   f. $D(t) = 60t - 2{,}000$
   g. $D(t) = 60t$
   h. $D(t) = 2{,}000 - 60t$
   j. $D(t) = 60 - 2{,}000t$
   k. $D(t) = 2{,}000 - 60t$

52. For \$2, a person can throw 3 balls at a dunk tank target in an attempt to dunk whoever is sitting on the platform. All proceeds from the dunk tank go to the police department. Of the 252 balls thrown at the target while Police Chief Hector Bailey was on the platform, only 12 resulted in a dunking. On average, how much money was spent for each dunking?
   a. \$21.00
   b. \$14.00
   c. \$16.80
   d. \$10.50
   e. \$0.66

53. In the following figure, which statement best describes the relationship between figure A and figure B?

   f. Figure B is a translation of figure A.

   g. Figure B is a reflection of figure A across a vertical line.

   h. Figure B is a reflection and translation of figure A across a vertical line.

   j. Figure B is a 60° clockwise rotation of figure A.

   k. Figure B is a 60° counterclockwise rotation of figure A.

54. The main fish tank at the East Point Aquarium is shaped like a rectangular prism. The tank is 16 feet deep, 24 feet wide, and 38 feet long. If the tank contains 4,660 cubic feet of water, how many more cubic feet of water must be added to completely fill the tank?

   a. 4,738 ft.^{3}

   b. 9,932 ft.^{3}

   c. 12,754 ft.^{3}

   d. 14,592 ft.^{3}

   e. 19,252 ft.^{3}

55. Which of the following answer choices represents the graph of $\frac{1}{2}x + y \le 3$?

   f.

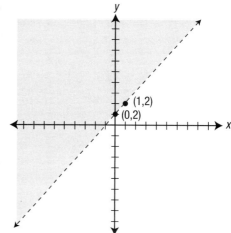

   g.

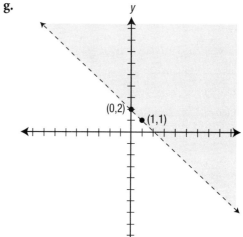

   h.

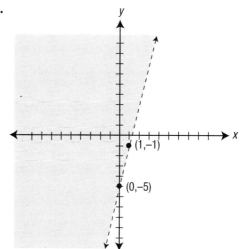

j.

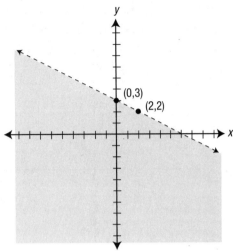

k. None of the above accurately represents the given inequality.

56.

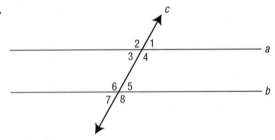

In the diagram, line c intersects lines a and b, which are parallel. If the measure of $\angle 4 = 122°$, what is $m\angle 7$?

a. 122°

b. 88°

c. 68°

d. 58°

e. It cannot be determined from the information given.

57. Which of the following ordered pairs is NOT a solution to the equation $y = 5x - 3$?

f. $(-6, -33)$

g. $(-2, -13)$

h. $(3, 12)$

j. $(5, 21)$

k. $(6, 27)$

58. Most cells are about 0.0025 cm in diameter. How is this measurement expressed in scientific notation?

a. 2.5×10^{-1} cm

b. 2.5×10^{-2} cm

c. 2.5×10^{-3} cm

d. 2.5×10^{3} cm

e. 2.5×10^{-4} cm

59. A cube with sides of length x centimeters has a surface area of $6x^2$ cm². If the length of each side of the cube was doubled, what would be the surface area of the resulting cube?

f. $12x^2$ cm²

g. $24x^2$ cm²

h. $36x^2$ cm²

j. $6(x + 2)^2$ cm²

k. $12(2x)^2$ cm²

60. A line passes through the points $(0, -1)$ and $(2, 3)$. What is the equation for the line?

a. $y = \frac{1}{2}x - 1$

b. $y = \frac{1}{2}x + 1$

c. $y = \frac{3}{2}x - 1$

d. $y = 2x + 1$

e. $y = 2x - 1$

61. Divide: $\frac{3x^3 - 6x^2 + 3x}{3x}$ $(x \neq 0)$

f. $x^2 - 2x$

g. $x^3 - 2x^2 + x$

h. $x^2 - 2x + x$

j. $x^2 - 2x + 1$

k. $x^3 - 2x$

62. Loretta bought a DVD player for $77. She bought the DVD player on sale for 30% off the original price. What was the original price of the DVD player?

a. $47

b. $107

c. $110

d. $100

e. $256

63. Find the area of the shaded region. Remember that the formula for the area of a circle is $A = \pi r^2$. Use 3.14 for π.

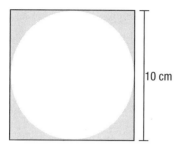

10 cm

    f. 100 cm^2
    g. 22.5 cm^2
    h. 78.5 cm^2
    j. 21.5 cm^2
    k. 178.5 cm^2

64. An empty cylindrical can has a height of 4 inches and a base with a radius of 3 inches. Melanie fills the can $\frac{2}{3}$ to the top with water. She then waters a plant with $\frac{1}{4}$ of the water in the can. What is the volume of the remaining water in the can? Use the following formula for calculating the volume of a cylinder: $V = \pi r^2(height)$.
    a. 24π in.^{3}
    b. 12π in.^{3}
    c. 18π in.^{3}
    d. 32π in.^{3}
    e. 30π in.^{3}

65. How much simple interest is earned on $300 deposited for 30 months in a savings account paying $7\frac{3}{4}\%$ simple interest annually? (The formula for calculating simple interest is Interest = Principal × Rate × Time, or I = PRT.)
    f. $5.81
    g. $12.62
    h. $19.76
    j. $23.25
    k. $58.13

66. If $\triangle ABC$, shown below, is reflected across the x-axis and then across the $y = x$ line to form $\triangle A'B'C'$, which point on the new triangle will have the greatest y-coordinate?

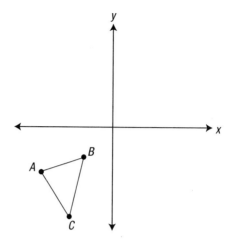

    a. A'
    b. B'
    c. C'
    d. A' and B' will have the same y-coordinate.
    e. It cannot be determined from the information given.

67. Which of the following equations represents the graph?

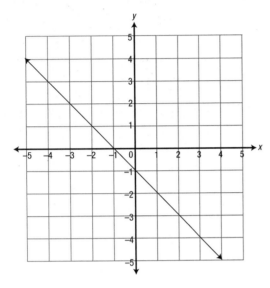

  f. $y = -x + 1$
  g. $y = x - 1$
  h. $y = -1 - x$
  j. $y = 2x - 2$
  k. $y = 1 + x$

68. At a President's Day sale, Yoko chose a sweater that was on sale for $16.20. The original price of the sweater was $36.00. What percent did Yoko save?
  a. 40%
  b. 45%
  c. 55%
  d. 58%
  e. 60%

69. As a person walks, the distance from the heel print of one foot to the heel print of the other foot is defined as one step, which averages about two feet. Renata would like to use the number of steps she takes when she walks to school to estimate the distance she travels. Which unit of measurement would be best to estimate this distance?
  f. millimeters
  g. centimeters
  h. meters
  j. decimeters
  k. kilometers

70. Stacey helped direct parking lot traffic at the school carnival. Between noon and 1:00 P.M. the number of cars doubled. In the next two hours, the number increased by two-thirds. Between 3:00 P.M. and 4:00 P.M., 37 cars drove away, leaving 123 cars in the lot. How many cars were in the lot at noon?
  a. 41
  b. 48
  c. 66
  d. 82
  e. 96

71. A pole that casts a 15-foot-long shadow stands near an 8-foot-high stop sign. If the shadow cast by the sign is 3 feet long, how high is the pole?
  f. 24 feet
  g. 28 feet
  h. 30 feet
  j. 40 feet
  k. 45 feet

72. Greg bought 15 raffle tickets for his girlfriend and $\frac{2}{3}$ as many raffle tickets for himself. There is one grand prize: a stereo system. If 75 raffle tickets were sold, what is the probability that either Greg or his girlfriend will win the stereo system?

  a. $\frac{10}{75}$
  b. $\frac{3}{25}$
  c. $\frac{50}{75}$
  d. $\frac{1}{8}$
  e. $\frac{1}{3}$

73. If Anthony's score was incorrectly reported as an 82 when his actual score on the test was a 90, which of the following statements would be true when his actual score is used in the calculations?

NAME	SCORE
Darin	95
Miguel	90
Anthony	82
Christopher	90
Samuel	88

  f. The mean, median, range, and mode will change.
  g. The mean, median, and range will change; the mode will remain the same.
  h. Only the mean and median will change.
  j. Nothing will change.
  k. none of the above

74. Quadrilateral *RSTU* in the diagram below is a parallelogram. What is the value of *x*?

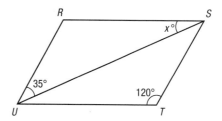

  a. 15
  b. 25
  c. 35
  d. 60
  e. 80

75. Zelda cuts a pizza in half in a straight line. She then cuts a line from the center to the edge, creating a 35-degree angle. What is the supplement of that angle?

  f. 55 degrees
  g. 145 degrees
  h. 35 degrees
  j. 70 degrees
  k. 180 degrees

76. Frank is a salesman for an appliance company. He earns a weekly salary of $500. In addition to his salary, Frank earns an 8% commission on the total cost of appliances he sells. If Frank sold $265,000 worth of appliances last year, which expression would give the total amount of money Frank earned last year?

  a. $500 + 0.08(265,000)$
  b. $52(500) + 0.08(265,000)$
  c. $\frac{265,000 + (0.8)500}{0.08}$
  d. $52(500) + 0.8(265,000)$
  e. $\frac{52(500) + 265,000}{0.08}$

77. Which of the answer choices represents the solution for the following systems of equations?

$$4x - 5y = 5$$
$$5y = 20 - x$$

f.

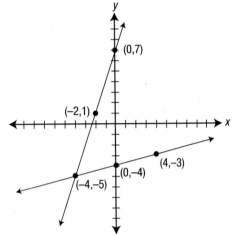

g.

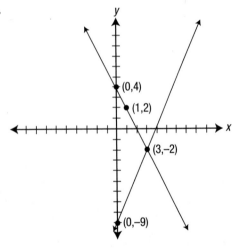

h.

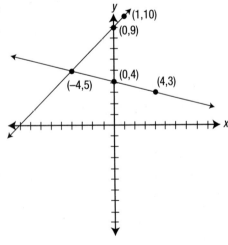

j.

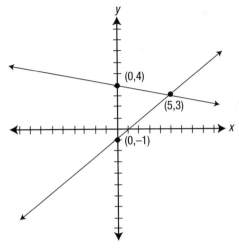

k. None of the above graphs models the system of equations accurately.

78. An organism initially contained 100 cells. The number of cells in the organism is tripling every 5 hours. The expression $100 \times 3^{\frac{h}{5}}$ can be used to calculate the number of cells in the organism after h hours. How many cells will the organism contain after 20 hours?

a. 1,200

b. 2,700

c. 1,500

d. 8,100

e. 12,000

79. If $\angle EDF$ and $\angle HIJ$ are supplementary angles, and $\angle SUV$ and $\angle EDF$ are also supplementary angles, then $\angle HIJ$ and $\angle SUV$ are

f. acute angles.

g. obtuse angles.

h. right angles.

j. congruent angles.

k. straight angles.

80. What is the number fourteen thousand written in scientific notation?

   a. 1.4×10^4

   b. 1.4×10^3

   c. $14{,}000$

   d. 140×10^2

   e. 14×10^3

81. If $x = 5$, $y = -1$, and $z = 2$, what is the value of $\dfrac{y^z - \frac{x}{y} + xyz}{xyz}$?

   f. -4

   g. $\frac{2}{5}$

   h. $\frac{3}{5}$

   j. $\frac{7}{5}$

   k. 6

82.

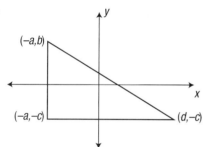

   What is the area of the triangle in the figure above?

   a. $\frac{1}{2} \times c \times a$

   b. $(c + b)(d + a)$

   c. $\frac{1}{2}(-a - c)(d + b)$

   d. $\frac{1}{2}(a + d)(b + c)$

   e. $\frac{1}{2}(d - a)(b - c)$

83. A surveyor is standing at point A and spots, in a direct line of sight, a building at point B 24 feet away. He also spots another building at point D, 48 feet from point B.

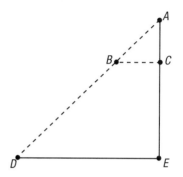

   If it is 60 feet from points D to E, how far is it from points B to C?

   f. 10 feet

   g. 12 feet

   h. 20 feet

   j. 28 feet

   k. 30 feet

84. At her party, Mackenzie put out a bowl containing 360 jellybeans. Marina came by and ate $\frac{1}{12}$ of the original total of jellybeans, Christina ate $\frac{1}{4}$ of the original total of jellybeans, Athena ate $\frac{1}{5}$ of the original total of jellybeans, and Jade ate $\frac{1}{8}$ of the original total of jellybeans. How many jellybeans were left?

   a. 120

   b. 240

   c. 237

   d. 123

   e. 124

85. $\left| (-2)^3 \right| - \left| -5 \times 2 \right| = ?$

   f. 2

   g. -2

   h. -4

   j. 18

   k. 16

86. If $\triangle ABC$ is rotated 90° clockwise about point A, the result is $\triangle A'B'C'$. What are the coordinates of $\triangle A'B'C'$?

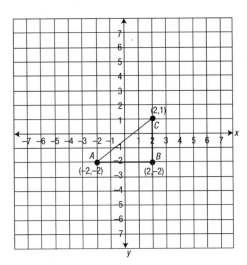

a. $(-2,-2), (-2,-6), (1,-6)$
b. $(-2,2), (-2,-2), (-5,-2)$
c. $(2,2), (2,6), (1,6)$
d. $(-2,-2), (-2,2), (-5,2)$
e. $(-2,-2), (-2,-2), (-5,-2)$

87. Dominick lies down on his back to stretch his legs. He keeps his left leg straight along the floor and raises his right leg in the air as high as he can. Which of the following is most likely the measure of the angle made by Dominick's right leg and the floor?
f. 10 degrees
g. 70 degrees
h. 180 degrees
j. 270 degrees
k. 360 degrees

88. $\angle A$, $\angle B$, and $\angle C$ meet the following conditions:
$\angle A$ and $\angle B$ are complementary.
$m\angle B = \frac{1}{2} m\angle C$.
$m\angle C = m\angle A + 30°$.
Which choice satisfies all three conditions?
a. $m\angle A = 40°$, $m\angle B = 50°$, $m\angle C = 100°$
b. $m\angle A = 40°$, $m\angle B = 140°$, $m\angle C = 10°$
c. $m\angle A = 50°$, $m\angle B = 40°$, $m\angle C = 25°$
d. $m\angle A = 50°$, $m\angle B = 40°$, $m\angle C = 100°$
e. $m\angle A = 50°$, $m\angle B = 40°$, $m\angle C = 80°$

89. What type of transformation will take the point $(-4,-5)$ and create the image $(-5,-4)$?
f. reflection over the x-axis
g. dilation of 2
h. reflection over the line $y = x$
j. translation of $T_{(1,1)}$
k. reflection over the y-axis

90. A jar contains b blue marbles, g green marbles, 20 yellow marbles, and r red marbles. If one marble is selected at random, what is the probability that it will be green?
a. $\frac{g}{b+g+r}$
b. $\frac{g}{b+20+r}$
c. $\frac{g}{b+20g+r}$
d. $\frac{g}{b+g+20+r}$
e. $\frac{g}{20bgr}$

91. Line *a* is parallel to line *b* in the figure below. What is the value of *x*?

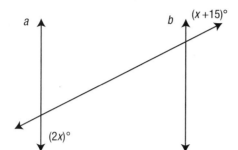

f. 15
g. 55
h. 65
j. 68
k. 106

92. Carol works part-time at the movie theater. Her schedule for the next three weeks lists the number of hours Carol will work each day. What is the median number of hours Carol will work over the next three weeks?

Sunday	Monday	Tuesday	Wednesday	Thursday	Friday	Saturday
0	7	0	4	4	5	0
0	4	6	5	3	2	0
0	5	4	3	6	5	0

a. 0 hours
b. 3 hours
c. 4 hours
d. 4.5 hours
e. 5 hours

93. If $\frac{3}{x+4} = \frac{5}{x}$, then $x -$
f. 10
g. −6
h. 0
j. 6
k. 10

94. A boy is flying a kite that has the shape of a parallelogram. The two posts keeping the kite together intersect at M. If $\overline{AM} = 5x - 10$ cm and $\overline{BM} = 2x + 50$ cm, what is the length of $\overline{AB}$?

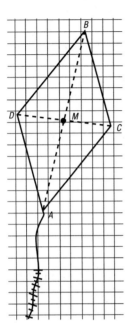

a. 180 cm

b. 90 cm

c. 120 cm

d. 100 cm

e. 20 cm

95. If $4g^2 + 15 = 16h^2 - 21$, what is the value of g in terms of h?

f. $(2h - 3)^2$

g. $(2h + 3)^2$

h. $\sqrt{4h^2 - 6}$

j. $\sqrt{4h^2 + 6}$

k. $(2h - 3)(2h + 3)$

Answers

Paragraph 1 (Q, U, R, T, S) Sentence **Q** refers to the classroom in the given sentence. Sentence **U** refers only to "it," but the classroom is the only singular noun that "it" could be. Each new sentence in this paragraph refers back to something in the preceding sentence.

Paragraph 2 (U, R, Q, T, S) Sentence **U** provides more information about Bob Kane, who is introduced in the given sentence. Sentence **R** tells why DC comics contacted Kane. Sentence **Q** refers back to Superman, so the correct order of those choices is **R, Q**. Sentences **T** and **S** add information about Kane's creation of Batman. Sentence **T** tells how Kane responded to DC's request. Sentence **S** tells where Kane's ideas may have come from.

Paragraph 3 (T, S, U, R, Q) Sentence **T** refers back to Kane's creation mentioned in the given sentence. Sentence **T** mentions the parents' death and sentence **S** notes the vow to avenge their deaths, so the order of these is logically **T, S**. Sentences **U** and **R** describe what Wayne did to fulfill his vow; the logical order is **U, R**. Sentence **Q** wraps up the paragraph.

Paragraph 4 (R, U, Q, T, S) Sentence **R** logically follows the given sentence, and these two sentences open the paragraph. The information about the pond freezing (**U**) must come between the memory of the pond (**R**) and the information about his daughter ice skating there (**Q**). Teaching his daughter to skate logically comes before remembering her skating. Sentence **S** gives an additional description of Abigail as she glides toward him.

Paragraph 5 (U, T, R, S, Q) Sentence **U** tells you that it is the first thing to keep in mind when evaluating a debate, so it follows the given sentence. Sentence **T** suggests what to consider "as you listen to the debate," and sentences **R** and **S** define the two attributes mentioned in sentence **T**. Sentence **Q** tells you what you should also consider, so it follows sentences **T, R**, and **S**.

6. a. Notice that we are not told in what direction Phil is driving—he might not be heading toward New York City at all. The only conclusion that we can draw is that he drove 14 hours over two days. The passage does not support any of the other answers.

7. f. The best way to solve this problem is to draw a diagram similar to the following:
  1. $S > D > K$
  2. $S > R > D$

Therefore, Sugar is the largest.

8. b. The only thing we know for sure is that Francesca sold more works than Mario. We do not know from the statements given to us whether Mario bought one of Francesca's works (choice **d**), which one of them is a better photographer (choice **c**), or whether one of them taught the other (choice **e**). The fact that they shared space for the exhibition means neither that they are the best photographers in New Orleans, nor that they necessarily live in New Orleans (choice **a**).

9. k. We are told in the first statement that the Smiths own two cars, so choice **f** is incorrect. Choices **h** and **j** could be true, but they are unsupported by the facts given us. Although we know that the Smiths have a small garage, that doesn't necessarily mean that it is a one-car garage (choice **g**). It could be a small two-car garage that will only accommodate the Jeep Cherokee, or Josh might prefer to park in the driveway. We cannot determine that any of these answers are true from the information given.

10. a. To answer this question, you need to determine what each letter represents. The words "attended the meeting" appear in all three sentences and the letters **A**, **E**, and **O** appear in all three sentences. Therefore, none of these letters can mean "Thirty." The word "Twenty" appears in the first and last sentences, and the letter **I** also appears in these two sentences. Therefore, **I** must represent "Twenty." The word "Thirty" appears in the second sentence only; the letter **Z** appears in the second sentence only. Therefore, **Z** must represent "Thirty."

11. j. In answering the preceding question, it was determined that the letters **A**, **E**, and **O** represent "attended the meeting," which means that choices **f** and **g** can be ruled out. Since the letter **U** appears in the first two sentences only, and the word "people" appears in the first two sentences only, then **U** must represent "people."

12. c. Megan states that "every rabbit that bit me was a gray rabbit," so she was not bitten by a white rabbit at the pet store. While choices **a** and **d** may or may not be true, there simply isn't enough information provided in Meghan's statement to correctly infer these claims. She does not say whether there were other gray animals at the pet store or whether she touched rabbits of any other color. There is no evidence to support choices **b** and **e**.

13. k. Since the first statement claims that all students in a creative writing program must submit a 50-page portfolio of original work, and we know that Veronika is in a creative writing program, we can conclude that Veronika submitted a 50-page portfolio of original work (choice **k**). Choice **f** cannot be true because 50 pages are required, not 30. We cannot conclude for certain that any of the other statements are true.

14. a. The statements tell us only that Roma could not get into his apartment (choice **a**). The fact that the lights in the kitchen were on does not mean that anybody was in the kitchen or in the apartment (choices **b** and **d**). Someone who lives in the apartment (including Roma) could have forgotten to turn the lights off when he or she left. Roma may have a spare set of keys at his office (choice **c**), but there is no evidence to support this choice.

15. k. Based solely on the information given, we do not know whether Anya would prefer not to have a full-time job (choice **g**), or whether she quit her job (choice **f**). Neither do we know whether Anya performed more than anyone else last February (choice **h**). Anya could be one of the best musicians in the city (choice **j**), but it is a statement unsupported by the facts. We do know from the first statement that Anya spends her time playing music all over New York City when she does not have a full-time job. Since we learned from the following statements that Anya does not perform when she works full-time and that she performed nearly every day last February, we can safely assume that she did not have a full-time job (choice **k**).

16. e. This choice includes the main points of the selection. Choice **b** is contradicted by the passage. The other choices are too narrow, covering only minor points from the passage.

17. f. Choice **f** is correct, according to the second sentence in paragraph 2. Choices **g**, **h**, and **k** are mentioned in the selection, but not as causing the islands. Choice **j** is not mentioned in the selection.

18. c. Paragraph 4 discusses the visitors to Acadia National Park, therefore, choice **c** is correct. There is no indication that choice **e** represents the author's opinion. Choices **a**, **b**, and **d** are not mentioned in the selection.

19. f. The first sentence in paragraph 3 states that the length of the Maine coastline is 2,500 miles. Paragraph 1 states that a straight-line distance between the northernmost and southernmost coastal cities—not the length of the coastline—is 225 miles, so **h** is incorrect. The remaining choices are also incorrect because they are not supported by the passage.

20. d. Choice **d** expresses the main idea of paragraph 4 of the selection. Choice **e** is just a detail from the paragraph. The information in choices **a**, **b**, and **c** is not expressed in paragraph 4.

21. e. This is the best choice because the answer to this question would be a summary of the passage. Choice **a** is too vague. The answer to choice **b** is a detail in the passage. The answer to choice **c** is not given—in fact, the last paragraph states that it will take years to understand the data that was collected. The answer to choice **d** is not given.

22. g. This is the correct choice based on information in the second paragraph. The paragraph tells what the scientists already knew, what they didn't know, and what they suspected, in other words, what their hypothesis was. The scientists then (choice **j**) waited for bad weather; (choice **h**) drove toward the storm front; (choices **f** and **k**) gathered storm data and measured water particles.

23. b. This is an inference drawn from the information in the fifth paragraph—the researchers say it will take years to understand the data they collected. The other choices are not supported by information in the passage.

24. k. This is clearly stated in the last sentence of the first paragraph. Choices **f** and **g** are wrong because these are caused by the storms; they are not the cause of the storms. There is no support for choices **h** and **j**.

25. c. See the fourth paragraph, which states that while they were in the air, they operated a special radar station "that would help them measure the shape and size of the water particles inside the clouds." Choice **a** may be true, but it is not the main reason for the plane ride. There is no support in the passage for the other choices.

26. c. The narrator clearly states that although she knows little of life beyond Lowood, she longs for freedom and is eager to go out into that world. Choices **a**, **d**, and **e** are contradicted by the passage. Choice **b** is only a detail from the passage.

27. g. The narrator mentions rules and systems, and contrasts that life with the wide and varied outside world. She says that the view from her window seemed a "prisonground." Lowood is both a structured and an isolated place. There is no evidence in the passage for any of the other choices.

28. a. The narrator "remembered that the real world was wide and awaited those who had courage to go forth." She also looks at the road from Lowood and longs to follow it. There is no evidence in the passage for any of the other choices.

29. k. Her desire for freedom and to explore the world are evident in this passage; she longs to follow the road that leads away from Lowood and she is "half desperate" for something new, something beyond Lowood and the rules and systems she has tired of. It is possible that she seems steady and sensible (choice **j**) but she makes it clear that she wants adventure. The passage does not support choices **f**, **g**, and **h**.

30. d. Lowood had been the narrator's home, and she says that all she knew was the school rules, duties, habits, faces, and so on. She had had no communication with the outside world. It is likely that she feels her prayer for freedom was unrealistic. The passage does not support any of the other choices.

31. b. This is the best choice because the passage consists mostly of the known differences between the Gunnison sage grouse and the northern sage grouse. Choices **a** and **e** are inferences that can be made based on the information in the passage; however, neither states what the passage is mostly about. Choice **c** is a detail from the passage. Choice **d** is incorrect.

32. j. This detail is supported by the information in the last paragraph. The other choices are refuted by information in the passage.

33. b. In the first paragraph, Gunnisons are described as having "been around at least as long as the northern sage grouse." Choice **a** is incorrect because this was the factor that caused the species to be officially recognized as its own species. Choice **c** refers to the last time a new species was added to the American Ornithological Union's official catalog of species. There is no support for choices **d** and **e**.

34. h. See the last sentence of the first paragraph. Choices **f** and **g** are incorrect because taxonomists believed the birds were the same species despite these differences. Choices **j** and **k** reference information presented in the passage not credited with proving the two species distinct.

35. a. The support for this choice is found in the second paragraph. For Dr. Young, the fact that the two species did not interbreed was a clue that they were separate species; therefore her research is based on this information. Choice **b** is too vague— neither of the species could recognize the other species' mating calls—at least that's what can be inferred since the sage grouse of each species would leave the area upon hearing the call of the other species. Choice **d** is incorrect. Choices **c** and **e** are not supported by the passage.

36. k. This is the correct choice supported by the information in the last paragraph. The northern sage grouse has a population of about 100,000 birds, and the "situation of the Gunnison sage grouse is even worse." Choice **f** is unsupported by the information in the passage. Choices **g**, **h**, and **j** are incorrect statements according to the passage.

37. b. This is the best choice because it correctly summarizes the main idea of the passage. Choices **a**, **c**, and **e** refer to facts taken from the passage; they are not main ideas. Choice **d** is not supported by the information in the text.

38. k. This detail is found in the second paragraph, where the passage explains that the lawyers sued the zookeepers under the Federal Animal Welfare Act, which says that zoos must safeguard primates' psychological well-being. There is no mention of not being fed (choice **f**), and although the lawyers did win a groundbreaking case (choice **g**), this was not why they sued the zookeepers. Choice **h** may be true, but the passage does not say this is why the lawyers sued. Choice **j** is wrong because the passage states that the lawyers heard about the chimps, not that they visited them.

39. d. This detail is taken from the closing sentence of the second paragraph. Choices **a**, **c**, and **e** are not supported by the information in the text. Choice **b** is supported by the text, but was not a direct result of the lawsuit.

40. h. This is an inference drawn from the information in the first three paragraphs. Through their actions, Barney and Judy's lawyers showed that they believe animals are more than mere property. Their actions show that they do believe animals deserve special protection (choice **f**), that they do have feelings (choice **g**), that they should be given the chance to socialize (choice **j**), and that they are creatures that need our help (choice **k**).

41. b. This is the only appropriate title. Choice **a** is too specific, since the passage indicates that making new friends is just one component of a book club. Choice **c** is incorrect because this passage does not contain numbered steps. Choice **e** is too vague, and the tone is inappropriate. Choice **d** is not found in the passage.

42. h. The second sentence of the second paragraph states this clearly. Choice **k** is not included in the passage and the none of the other choices is the first thing to be done.

43. e. Note that the question asks what would not be covered. Deciding on the club's focus— the kinds of books or genre the club will read—should be done prior to this meeting and prior to recruiting members, according to the second paragraph.

44. f. Note that the question asks what should not be done. The passage states this is one possible focus, but does not say successful book clubs must focus exclusively on one genre. The other choices are all in the passage. Choice **h** might seem attractive at first, but the passage clearly states that a focus should be chosen, even if that focus is defined as flexible and open.

45. d. The tone and specificity of the passage infer that a successful book club requires careful planning.

46. c. 8. List the number in order of smallest to greatest and average the center two numbers: 4, 6, 8, 8, 8, 9, 9, 12.

47. g. a line with slope $= \frac{1}{2}$. The equation $2y + 10 = x$ can be written in slope-intercept form as $y = \frac{x}{2} - 5$, which has a slope of $\frac{1}{2}$.

48. b. $-4x^2 - 5x + 8$. Add all the same terms, but with their opposite signs in order for the trinomial to equal to 0.

49. j. $2x^2 + 10x + 8$. $A = 2x^2$; $B = 2x$; $C = 8x$; $D = 8$; and their sum is $2x^2 + 10x + 8$.

50. c. (0,1). Divide by 4 to get $y = 3x + 1$, where the y-intercept is 1.

51. k. $D(t) = 2{,}000 - 60t$. The starting distance from Chicago to Los Angeles is 2,000 miles, so that comes first in the equation. Then, for every hour, the train gets 60 miles closer to LA, so subtract $60t$ from 2,000 to create the correct equation.

52. b. $14.00. $\frac{252}{3} = 84$ shows how many $2 packages were purchased. Since 84 chances were purchased at $2 each, $168 was earned. To find the average amount of money spent for each dunking, divide $168 by 12 to get $14.

53. g. Figure B is a reflection of figure A across a vertical line. Figure A is a mirror image of figure B across a vertical line, so it is a vertical reflection.

54. b. 9,932 ft.^{3}. The tank's volume is (16 feet)(24 feet)(38 feet) = 14,592 ft^3. 14,592 ft^3 – 4,660 ft^3 = 9,932 ft^3.

55. j.

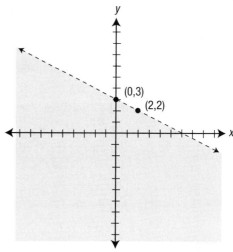

$y \leq -\frac{1}{2}x + 3$ has a slope of $-\frac{1}{2}$ and a y-intercept of 3 and the graph must be shaded under a solid line since y is less than or equal to $-\frac{1}{2}x + 3$.

56. d. 58°. $\angle 4$ and $\angle 8$ are corresponding angles, so $m\angle 8 = 122°$. Since $\angle 7$ is supplementary to $\angle 8$, $m\angle 7 = 58°$.

57. j. (5,21). $y = 5(5) - 3 = 25 - 3 = 22$, not 21, so $(5, 21)$ does not work.

58. c. 2.5×10^{-3} cm. The decimal in 0.0025 cm needs to move 3 times to the right in order to put it in scientific notation.

59. g. $24x^2$ cm^2. A cube with sides of length $(2x)$ will have a single-face area of $4x^2$ and a surface area of $6(4x^2)$ cm^2. $6(4x^2)$ cm$^2 = 24x^2$ cm^2.

60. e. $y = 2x - 1$.
First find the slope using $\frac{3-(-1)}{2-0} = \frac{4}{2} = 2 = m$.
Then notice that $(0,-1)$ defines the y-intercept at -1.
So use $y = mx + b$ and sub in $m = 2$ and $b = -1$ to get $y = 2x + (-1)$.

61. j. $x^2 - 2x + 1$. Divide each individual term in the numerator by $3x$ and subtract the exponents of like bases to get: $x^2 - 2x + 1$.

62. c. **$110.** If Loretta bought the DVD player for 30% off, that means that she only paid 70% of the original price: (70%)(original price) = $77. Divide $77 by 0.70 to get $110 as the original price.

63. j. **21.5 cm².** Find the area of the circle and subtract that from the area of the square in order to find the area of the shaded region. Area of the circle = $\pi r^2 = \pi 5^2 = 3.14(25) = 78.50$.
Area of the square = 100.
Area of shaded region = $100 - 78.5 = 21.5$.

64. c. **18π in.³** The volume of the cylindrical can will be the (area of the bottom circle) times (cylinder's height): $(\pi r^2)(\text{height}) = (\pi 3^2)(4) = 36\pi$ in.³ is the total volume of the can. Filling it $\frac{2}{3}$ to the top will be 24π in.³ Watering the plant with $\frac{1}{4}$ of the water will use 6π in.³, so there will be 18π in.³ left.

65. k. **$58.13.** $300 is the principal. The interest rate of $7\frac{3}{4}\%$, must be written as a decimal as 0.0775. Thirty months needs to be converted into years by dividing it by 12, since t is always in years. Interest = $300 \times 0.0775 \times \frac{30}{12} = \$58.125$.

66. b. **B'.** After reflecting ΔABC across the x-axis, the point with the largest x-coordinate will have the largest y-coordinate after the triangle is reflected across the $y = x$ line to form $\Delta A'B'C'$. This is because the x- and y-coordinates swap places.

67. h. **$y = -1 - x$.** The line has a slope of -1 and a y-intercept of -1.

68. c. **55%.** The formula for percent decrease is $\frac{\text{difference}}{\text{original}}$. The difference between the original and sale price was $19.80. Therefore, $\frac{\$19.80}{36} = 55\%$.

69. h. **meters.** Two feet is approximately $\frac{2}{3}$ of a meter, so meters would be the best measure for Renata's distance to school.

70. b. **48.** Assume that there are c cars in the lot at noon. When the cars double this will be represented by $2c$. When $\frac{2}{3}$ more cars come, that will be represented by $(2c)(\frac{2}{3})$. An equation can be written: $2c + 2c(\frac{2}{3}) - 37 = 123$. Solving for c results in 48 cars.

71. j. **40 feet.** This problem is modeled with two similar right triangles: The smaller one has a height of 8 and a shadow of 3. The larger one has a shadow of 15 and a height of h. Setting up a proportion gives:
$\frac{height}{shadow} = \frac{8}{3} = \frac{h}{15}$
$3h = 120$
$h = 40$

72. e. **$\frac{1}{3}$.** Greg bought $(\frac{2}{3})(15) = 10$ tickets for himself and 15 for his girlfriend, which is 25 tickets in total. 25 out of the 75 raffle tickets sold, gives a probability of $\frac{25}{75}$ or $\frac{1}{3}$ that either of them will win.

73. k. **none of the above.** The mean and range would change. The median and mode would still be the same.

74. b. **25.** In quadrilaterals, adjacent angles are supplementary, so because $m < T = 120°$, $\angle RST = 60°$. Since $RSTU$ is a parallelogram, $\angle UST$ must be 35°. $x = 60° - 35° = 25°$.

75. g. **145 degrees.** The supplement of an angle creates a sum of 180°. $35° + 145° = 180°$.

76. b. **$52(500) + 0.08(265,000)$.** Multiply his sales by 8% in decimal form and then add that to his weekly salary multiplied by 52 weeks in the year: $(\$265,000)(0.08) + (52 \text{ weeks})(\$500 \text{ per week})$.

77. j.

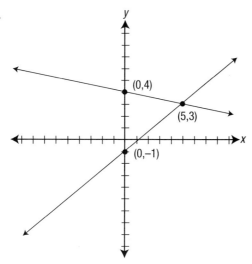

$4x - 5y = 5$ written in slope-intercept form is $y = \frac{4}{5}x - 1$, which has a slope of $\frac{4}{5}$ and a y-intercept of -1. $5y = 20 - x$ written in slope-intercept form is $y = -\frac{1}{5}x + 4$, which has a slope of $-\frac{1}{5}$ and a y-intercept of 4.

78. d. 8,100. Substitute 5 hours in for x and compute: $100 \times 3^{\frac{20}{5}} = 100 \times 3^4 = 100 \times 81 = 8,100$.

79. j. congruent angles. If $\angle EDF + \angle HIJ = 180° = \angle EDF + \angle SUV$, then $m\angle HIJ$ must equal $m\angle SUV$.

80. a. 1.4×10^4. In scientific notation, the first term must be greater than or equal to one and less than 10. In this case, 1.4 needs the decimal to move four places to the right in order to add on the three zeros.

81. g. $\frac{2}{5}$.

$$\frac{y^2 - \frac{x}{y} + xyz}{xyz} = \frac{(-1)^2 - \frac{(5)}{(-1)} + 5(-1)(2)}{5(-1)(2)} = \frac{1 + 5 - 10}{-10} = \frac{-4}{-10} = \frac{2}{5}$$

82. d. $\frac{1}{2}(a + d)(b + c)$. Subtract the x-coordinates to determine the length of the base. Subtract the y-coordinates to determine the height. Area $= \frac{1}{2}$(base)(height) $= \frac{1}{2}(d - (-a))(b - (-c)) = \frac{1}{2}(a + d)(b + c)$.

83. h. 20 feet. This is a similar triangle problem. Since the distance from A to D is 72 feet (48 + 24), it can be determined that ΔAED is three times as large as ΔACB (since $24 \times 3 = 72$ and those are corresponding parts). Therefore, the distance from D to E will be three times as long as the distance from B to C. The distance from B to C must be 20 feet since the distance from D to E is 60 feet.

84. d. 123. Marina ate $\frac{1}{12}$ of 360 which is $\frac{360}{12} = 30$. Christina ate $\frac{1}{4}$ of 360 which is $\frac{360}{4} = 90$. Athena ate $\frac{1}{5}$ of 360 which is $\frac{360}{5} = 72$. Jade ate $\frac{1}{8}$ of 360 which is $\frac{360}{8} = 45$. Together they all ate 237 jellybeans so there were 123 left.

85. g. –2. Do the calculations inside the absolute values first, and then change them to positive numbers as you remove the absolute value sign. Remember, unlike with parentheses, a minus does not combine with a negative inside an absolute value sign to form a plus. $|(-2)^3| - |-5 \times 2| = |-8| - |-10| = 8 - 10 = -2$.

86. a. (–2,–2), (–2,–6), (1,–6). Point A will stay the same, so that narrows your choices down to **a**, **d**, and **e**. Point B will be 4 vertical positions down from point A after being rotated 90° clockwise around $(-2,2)$, which will put point B at $(-2,-6)$.

87. g. 70 degrees. Lifting one's leg past 120 degrees is not likely and 10 degrees is too little to be realistic. 70 degrees is the only likely option.

88. e. $m\angle A = 50°$, $m\angle B = 40°$, $m\angle C = 80°$. Only **a** and **e** satisfy the second condition, $m\angle B = \frac{1}{2}m\angle C$. Of those, only choice **e** satisfies the third condition.

89. h. reflection over the line $y = x$. Reflecting over the $y = x$ line will switch the coordinates but not the signs.

90. d. $\frac{g}{b+g+20+r}$. The probability that a desired event will happen is $\frac{\text{number of desired choices}}{\text{total number of choices}}$. In this case, the total number of choices is $b + g + 20 + r$ and the number of green marbles is g.

91. g. 55. Since line a is parallel to line b, the two given angles are supplementary. Therefore $2x + (x + 15) = 180$. $3x = 165$ so $x = 55$.

92. c. 4 hours. There are 21 days in 3 weeks and the median work shift will be the central, or 11th term. When putting Carol's hours in order of least to greatest, 4 hours is the 11th term.

93. f. −10. Using cross multiplication, $5(x + 4) = 3x$, which simplifies to $5x + 20 = 3x$ and then to $20 = -2x$. Therefore, $x = -10$.

94. a. 180 cm. The diagonals of a kite bisect one another, so the length of segment AM = length of segment BM. $5x - 10 = 2x + 50$ which simplifies to $3x = 60$, and then $x = 20$. $[5(20) - 10] + [2(20) + 50] = 180$.

95. k. $(2h - 3)(2h + 3)$. Starting with $4g^2 + 15 = 16h^2 - 21$, subtract 15 from both sides to get $4g^2 = 16h^2 - 36$. Then divide by 4 to get $g^2 = 4h^2 - 9$. This is a difference of perfect squares, so it can be factored as follows: $(a^2 - b^2) = (a - b)(a + b)$. Therefore, $4h^2 - 9 = (2h - 3)(2h + 3)$.

C H A P T E R

7 ▶ PRACTICE TEST 5

T he *SHSAT Power Practice* tests will help you prepare for the high-stakes exams given to students applying for New York City's specialized high schools. Each practice test consists of sample questions like
those you will find on the official SHSAT.

The 45-question verbal section and 50-question math section were developed by education experts. These
tests will show you how much you know and what kinds of problems you still need to study. Mastering these
practice tests will allow you to reach your highest potential on the real SHSAT.

PART I VERBAL

Scrambled Paragraphs

Paragraph 1

(q) (r) (s) (t) (u)
(q) (r) (s) (t) (u)
(q) (r) (s) (t) (u)
(q) (r) (s) (t) (u)
(q) (r) (s) (t) (u)

Paragraph 2

(q) (r) (s) (t) (u)
(q) (r) (s) (t) (u)
(q) (r) (s) (t) (u)
(q) (r) (s) (t) (u)
(q) (r) (s) (t) (u)

Paragraph 3

(q) (r) (s) (t) (u)
(q) (r) (s) (t) (u)
(q) (r) (s) (t) (u)
(q) (r) (s) (t) (u)
(q) (r) (s) (t) (u)

Paragraph 4

(q) (r) (s) (t) (u)
(q) (r) (s) (t) (u)
(q) (r) (s) (t) (u)
(q) (r) (s) (t) (u)
(q) (r) (s) (t) (u)

Paragraph 5

(q) (r) (s) (t) (u)
(q) (r) (s) (t) (u)
(q) (r) (s) (t) (u)
(q) (r) (s) (t) (u)
(q) (r) (s) (t) (u)

Logical Reasoning

6. (a) (b) (c) (d) (e)
7. (f) (g) (h) (j) (k)
8. (a) (b) (c) (d) (e)
9. (f) (g) (h) (j) (k)
10. (a) (b) (c) (d) (e)
11. (f) (g) (h) (j) (k)
12. (a) (b) (c) (d) (e)
13. (f) (g) (h) (j) (k)
14. (a) (b) (c) (d) (e)
15. (f) (g) (h) (j) (k)

Reading

16. (a) (b) (c) (d) (e)
17. (f) (g) (h) (j) (k)
18. (a) (b) (c) (d) (e)
19. (f) (g) (h) (j) (k)
20. (a) (b) (c) (d) (e)
21. (f) (g) (h) (j) (k)
22. (a) (b) (c) (d) (e)
23. (f) (g) (h) (j) (k)
24. (a) (b) (c) (d) (e)
25. (f) (g) (h) (j) (k)
26. (a) (b) (c) (d) (e)
27. (f) (g) (h) (j) (k)
28. (a) (b) (c) (d) (e)
29. (f) (g) (h) (j) (k)
30. (a) (b) (c) (d) (e)

31. (f) (g) (h) (j) (k)
32. (a) (b) (c) (d) (e)
33. (f) (g) (h) (j) (k)
34. (a) (b) (c) (d) (e)
35. (f) (g) (h) (j) (k)
36. (a) (b) (c) (d) (e)
37. (f) (g) (h) (j) (k)
38. (a) (b) (c) (d) (e)
39. (f) (g) (h) (j) (k)
40. (a) (b) (c) (d) (e)
41. (f) (g) (h) (j) (k)
42. (a) (b) (c) (d) (e)
43. (f) (g) (h) (j) (k)
44. (a) (b) (c) (d) (e)
45. (f) (g) (h) (j) (k)

PART II MATHEMATICS

46. (a) (b) (c) (d) (e)
47. (f) (g) (h) (j) (k)
48. (a) (b) (c) (d) (e)
49. (f) (g) (h) (j) (k)
50. (a) (b) (c) (d) (e)
51. (f) (g) (h) (j) (k)
52. (a) (b) (c) (d) (e)
53. (f) (g) (h) (j) (k)
54. (a) (b) (c) (d) (e)
55. (f) (g) (h) (j) (k)
56. (a) (b) (c) (d) (e)
57. (f) (g) (h) (j) (k)
58. (a) (b) (c) (d) (e)
59. (f) (g) (h) (j) (k)
60. (a) (b) (c) (d) (e)
61. (f) (g) (h) (j) (k)
62. (a) (b) (c) (d) (e)

63. (f) (g) (h) (j) (k)
64. (a) (b) (c) (d) (e)
65. (f) (g) (h) (j) (k)
66. (a) (b) (c) (d) (e)
67. (f) (g) (h) (j) (k)
68. (a) (b) (c) (d) (e)
69. (f) (g) (h) (j) (k)
70. (a) (b) (c) (d) (c)
71. (f) (g) (h) (j) (k)
72. (a) (b) (c) (d) (e)
73. (f) (g) (h) (j) (k)
74. (a) (b) (c) (d) (e)
75. (f) (g) (h) (j) (k)
76. (a) (b) (c) (d) (e)
77. (f) (g) (h) (j) (k)
78. (a) (b) (c) (d) (e)
79. (f) (g) (h) (j) (k)

80. (a) (b) (c) (d) (e)
81. (f) (g) (h) (j) (k)
82. (a) (b) (c) (d) (e)
83. (f) (g) (h) (j) (k)
84. (a) (b) (c) (d) (e)
85. (f) (g) (h) (j) (k)
86. (a) (b) (c) (d) (e)
87. (f) (g) (h) (j) (k)
88. (a) (b) (c) (d) (e)
89. (f) (g) (h) (j) (k)
90. (a) (b) (c) (d) (e)
91. (f) (g) (h) (j) (k)
92. (a) (b) (c) (d) (e)
93. (f) (g) (h) (j) (k)
94. (a) (b) (c) (d) (e)
95. (f) (g) (h) (j) (k)

Part 1—Verbal

The Verbal Test includes 45 questions in these three sections:

- Scrambled Paragraphs, 5 paragraphs (each counts double)
- Logical Reasoning, 10 questions, numbered 6–15
- Reading, 30 questions, numbered 16–45

Scrambled Paragraphs

This section tests your ability to organize a paragraph well. There are five paragraphs, presented in scrambled order. Your job is to put them in the best order to make a clear, coherent paragraph. Each correct answer counts double; these five paragraphs are worth 10 points out of the 50-point verbal test.

The first sentence in each paragraph is given. The remaining five sentences are listed in random order. Read each group of sentences carefully, and then decide on the best arrangement for them. Use the blanks at the left of each sentence to number these sentences from 1 to 5, showing the order they should be in.

Paragraph 1

In the damp, swampy bogs of coastal North Carolina grows one of nature's most interesting creations: the Venus flytrap.

_____ **Q.** This unique plant has a one-foot-high stalk that is topped by a pair of shiny, oddly shaped leaves with spiny bristles on their edges.

_____ **R.** After about two weeks, when the insect has been completely digested, the leaves open once again to await another victim.

_____ **S.** When insects land on the leaves of this plant, however, they get an unpleasant surprise.

_____ **T.** Insects are attracted to these shiny, bristly leaves.

_____ **U.** The leaves snap shut, trapping the insect and providing the Venus flytrap with its next meal.

Paragraph 2

The stones known as *beach agates* are a form of quartz.

_____ **Q.** They were formed by water-borne silicones, oxides, and metals that were deposited in basalt and other earth forms.

_____ **R.** Not only do they come in many colors, these stones are very hard and can be highly polished.

_____ **S.** Thousands of years before the Ice Age, these beach agates formed in gravel beds along the coastal plains.

_____ **T.** Colorful, polished agates have long been used by artists and craftspeople to make beads, brooches, and other ornaments.

_____ **U.** Because of this variety of oxides and minerals, agates can be found in a multitude of colors.

Paragraph 3

Charles Darwin was a biologist whose famous theory of evolution influenced the ideas of many scientists and philosophers.

_____ **Q.** Those creatures that survive pass on favorable changes to their offspring.

_____ **R.** After years of careful study, Darwin attempted to show that higher species come into existence as a result of the gradual transformation of lower species.

_____ **S.** Those creatures that embody favorable changes survive to produce the next generation.

_____ **T.** Darwin called this process the "survival of the fittest."

_____ **U.** He believed that this process of transformation could be explained through the effect of the natural environment upon living creatures.

Paragraph 4

Today, painters often use their photographs to document the scenes they later will paint.

_____ **Q.** The camera also records the light and colors just as they were at the time that the artist chose the scene.

_____ **R.** A sketch may be done first, recording the composition as the artist views it and responds to it.

_____ **S.** Between the sketch and the photo, artists have just about all they will need to begin the painting.

_____ **T.** Then the camera records the details and fills in whatever the sketch may have missed.

_____ **U.** Photographs function most effectively when they are used along with the artist's sketch of the same scene.

Paragraph 5

Effective face-to-face communication depends on the ability to listen well.

_____ **Q.** Unfortunately, many of us hear what others say without really listening to the message they are sending.

_____ **R.** You can also respond actively by leaning forward and looking the speaker in the eye to show that you are paying attention.

_____ **S.** The second step is to actively respond to what you hear.

_____ **T.** The first step to better listening is to pay attention, which means no

doodling, daydreaming, or checking your watch.

_____ **U.** For example, use nonverbal responses, nod or shake your head, laugh or smile, and make other appropriate gestures.

Logical Reasoning

The questions in this section test your ability to reason well, that is, to figure out what the facts you know can or can't possibly mean. Read the statements carefully, then choose the best answer based *only* on the information given. Note carefully the words used in each question. For example, one thing can be larg*er* than another without being the larg*est* in the group. In answering some of these questions, it may be useful to draw a rough diagram or make a list that gives real world values to the information.

This section contains 10 questions, numbered 6–15.

6. For his drawings, Christopher uses only his special black-ink pen. Tonight, Christopher has forgotten his pen at a friend's apartment.

Which of the following can we conclude is true?

a. Christopher is usually very forgetful.
b. Christopher does not have too many friends.
c. Christopher will probably not draw tonight.
d. Christopher draws every night.
e. Christopher has decided to use a different pen.

7. The alarm goes off at the State National Bank. Officer Mancini is patrolling in her squad car ten miles away. Officer Fromme is patrolling five miles away, Officer Smith, seven miles. Officer Sexton is farther away than Fromme, but closer than Smith.

Approximately how far away from the bank is Sexton?

f. nine miles

g. eight miles

h. six miles

j. four miles

k. It cannot be determined from the information given.

8. When it is cold and rainy outside, Dan does not like to go out. When the weather is nice, Dan never stays in. Yesterday, Dan spent an entire day at home.

Which of the following must be true?

a. Dan likes to stay at home and read.

b. Dan does very little work at home.

c. Yesterday, the weather was nice.

d. Yesterday was a cold and rainy day.

e. It cannot be determined from the information given.

9. Read the following statements:

■ During the past year, Zoe read more books than Jenna.

■ Jenna read fewer books than Heather.

■ Heather read more books than Zoe.
If the first two statements are true, the third is

f. true.

g. false.

h. probably false.

j. partly true.

k. not possible to determine from the information given.

10. Read the following information, then answer the question.

■ The hotel is two blocks east of the drugstore.

■ The market is one block west of the hotel.

■ The post office is one block north of the market.

Where is the post office in relation to the drugstore?

a. The post office is northwest of the drugstore.

b. The post office is northeast of the drugstore.

c. The post office is southwest of the drugstore.

d. The post office is southeast of the drugstore.

e. It cannot be determined from the information given.

11. Jenny is flying nonstop from New York City to Vancouver, BC, via the Cathay Pacific airline. The nonstop flight from New York City to Vancouver, BC, takes approximately six hours.

Which of the following can we conclude is true?

f. Jenny prefers flights between the cities where she can fly nonstop.

g. Jenny will arrive to Vancouver, BC, in approximately six hours.

h. The Cathay Pacific airline serves better food than all other airlines known to Jenny.

j. Jenny prefers to fly to Vancouver, BC, from New York, rather than from New Jersey.

k. It takes approximately six hours by plane to get to Vancouver, BC, from New York City.

12. Georgia is older than her cousin Marsha. Marsha's brother Bart is older than Georgia. When Marsha and Bart visit Georgia, all three like to play Monopoly. Marsha wins more often than Georgia.

Based only on the information above, which of the following statements must be true?

a. When he plays Monopoly with Marsha and Georgia, Bart often loses.
b. Of the three, Georgia is the oldest.
c. Georgia hates to lose at Monopoly.
d. The youngest Monopoly player wins the most games.
e. Of the three, Marsha is the youngest.

13. The Petersens have three children—one-year-old Pete, three-year-old Jim, and five-year-old Michele. Linda is six years old.

Given these facts, we can conclude that

f. Linda is younger than Michele.
g. Linda is Michele's friend.
h. Linda is the Petersens' neighbors' daughter.
j. Linda is not one of the Petersens' children.
k. Linda is a cousin to the Petersen's children.

14. Mr. Magnani is a professor of film who frequently reads lectures on the life and works of Federico Fellini. Fellini is Mr. Magnani's favorite filmmaker. Like Fellini, Mr. Magnani was born and raised in Italy.

Which of the following can we assume is true?

a. Federico Fellini is one of the greatest filmmakers of our time.
b. Mr. Magnani was once a filmmaker himself.
c. Mr. Magnani is very knowledgeable about many things besides film.
d. Federico Fellini was born and raised in Italy.
e. Mr. Magnani and Federico Fellini met in Italy.

15. In a four-day period—Monday through Thursday—each of the following temporary office workers worked only one day, each a different day. Ms. Johnson was scheduled to work on Monday, but she traded with Mr. Carter, who was originally scheduled to work on Wednesday. Ms. Falk traded with Mr. Kirk, who was originally scheduled to work on Thursday.

After all the switching was done, who worked on Tuesday?

f. Mr. Carter
g. Ms. Falk
h. Ms. Johnson
j. Mr. Kirk
k. cannot be determined from the information given

Reading Comprehension

This section tests your reading comprehension—your ability to understand what you read. Read each passage carefully and answer the questions that follow it. If necessary, you can reread the passage to be certain of your answers. Remember that your answers must be based only on information that is actually in the passage.

Read the following passage, then answer Questions 16 through 22.

The lives of the ancient Greeks revolved around *eris*, a concept by which they defined the universe. They believed that the world existed in a condition of opposites. If there was good, then there was evil; if there was love, then there was hatred; joy, then sorrow; war, then peace; and so on. The Greeks believed that good *eris* occurred when one held a balanced outlook on life and coped with problems as they arose. It was a kind of ease of living that came from trying to bring together the great opposing

forces in nature. Bad *eris* was evident in the
violent conditions that ruled men's lives.
Although these things were found in nature
and sometimes could not be controlled, it was
believed that bad *eris* occurred when one
ignored a problem, letting it grow larger until it
destroyed not only that person, but his family
as well. The Greeks saw *eris* as a goddess: Eris,
the Goddess of Discord, better known as
Trouble.

One myth that expresses this concept of
bad *eris* deals with the marriage of King Peleus
and the river goddess Thetis. Zeus, the supreme
ruler, learns that Thetis would bear a child
strong enough to destroy its father. Not
wanting to father his own ruin, Zeus convinces
Thetis to marry a human, a mortal whose child
could never challenge the gods. He promises
her, among other things, the greatest wedding
in all of Heaven and Earth and allows the
couple to invite whomever they please.

This is one of the first mixed marriages of
Greek mythology and the lesson learned from it
still applies today. Thetis and Peleus do invite
everyone—except Eris, the Goddess of Discord.
In other words, instead of facing the problems
brought on by a mixed marriage, they turn their
backs on them. They refuse to deal directly with
their problems and the result is tragic.

In her fury, Eris arrives, ruins the
wedding, causes a jealous feud between the
three major goddesses over a golden apple, and
sets in place the conditions that lead to the
Trojan War. The war would take place 20 years
in the future, but it would result in the death of
the only child of the bride and groom, Achilles.
Eris would destroy the parents' hopes for their
future, leaving the couple with no legitimate
heirs to the throne. Hence, when we are told,
"If you don't invite trouble, trouble comes," it
means that if we don't deal with our problems,
our problems will deal with us—with a
vengeance! It is easy to see why the Greeks

considered many of their myths learning
myths, for this one teaches us the best way to
defeat that which can destroy us.

16. Which of the following best tells what this
passage is about?
   a. the wedding of a goddess and a mortal
   b. the goddess Eris, who brought great benefits
to the ancient Greek culture
   c. the concept of opposites by which the
ancient Greeks defined the universe
   d. the tragic death of Achilles in the Trojan
War
   e. King Peleus and the river goddess Thetis

17. According to the passage, the ancient Greeks
believed that the concept of eris defined the
world as
   f. a hostile, violent place.
   g. a condition of opposites.
   h. a series of problems.
   j. a mixture of gods and man.
   k. a place where only gods should live.

18. Most specifically, *bad eris* is defined in the
passage as
   a. the violent conditions of life.
   b. the problems man encounters.
   c. the evil goddess who has a golden apple.
   d. the murderer of generations.
   e. the doom of future generations.

19. It can be inferred that Zeus convinced Thetis to
marry Peleus because
   f. he needed to buy the loyalty of a great king
of mankind.
   g. he feared the gods would create bad *eris* by
competing over her.
   h. he feared the Trojan War would be fought
over her.
   j. he feared having an affair with her and,
subsequently, a child by her.
   k. he feared his wife's jealousy.

20. It can also be inferred that Zeus did not fear a child sired by King Peleus because

    a. he knew that the child could not climb Mt. Olympus.

    b. he knew that the child would be killed in the Trojan War.

    c. he knew that no matter how strong the child of a mortal might become, he couldn't overthrow an immortal god.

    d. he knew that Thetis would always love him above everyone else.

    e. Thetis would have greatest wedding in all of Heaven and Earth.

21. According to the passage, Achilles

    f. defeated Zeus during the Trojan War.

    g. would die during the Trojan War.

    h. was born 20 years after the war because of the disruption Eris caused at the wedding.

    j. was not really the son of Peleus.

    k. was the son of Zeus.

22. Which of the following statements is the message offered in this myth?

    a. Do not consider a mixed marriage.

    b. Do not anger the gods.

    c. Do not engage in wars.

    d. Do not take myths seriously.

    e. Do not ignore the problems that arise in life.

Read the following, then answer questions 23 through 26.

Communications Workshop Overview
Welcome to the Communication Workshop. At today's conference, we'll be discussing many aspects of business communication. You'll learn many useful strategies you can use to make your workplace discussions effective and productive. This list presents some key concepts we'll discuss during the seminar.

1. Listen, Listen, Listen
You have to listen before you can respond. It sounds simple, but you'd be surprised how often people jump into a conversation without really listening to what others are saying. Focus on what the other person is saying, not what you want to say in response. Another useful strategy is to summarize what you just heard a speaker say before adding your own comments. This technique can help you correct any misunderstandings, too.

2. Use Professional Language
Remember that a workplace is a professional setting. No matter how strongly you feel about a subject, you should choose your words carefully. Some words and phrases that might be suitable at home are not appropriate at work.

3. Be Specific
The more specific you are, the more clearly you communicate. Instead of saying "The quarterly report is a mess," take the time to describe the exact problems that you have identified. A general statement is much more likely to be misinterpreted or misunderstood than a specific one.

4. Avoid Negative Statements
You can think of this as the *No "No" Rule*. Work is often about problem solving, but too many negative statements can create a gloomy and pessimistic work atmosphere. You'll get much better reactions if you balance negative statements with positive ones. You might feel like shouting "There's no way we'll ever meet this deadline!" but you'll help your team a lot more if you say "Last time we had a deadline like this it seemed impossible, but we pulled together and got the job done!"

5. Use Notes
Notes can help you keep your business communications focused and productive. Before a meeting, jot down a list of the topics

you want to cover or questions you want to ask. During or after a meeting, write a quick summary that covers the main points discussed. You can e-mail your notes to everyone who participated in the discussion, so that you all have the same record.

23. Which of the following best tells what this passage is about?
    a. a conference about making decisions based on effective communication
    b. a conference about using notes to improve your business communications
    c. a conference about avoiding negative statements to create a positive work environment
    d. a conference about using communication skills to manipulate consumers
    e. a conference about ways to improve your business communication skills

24. According to the overview, which of the following will NOT be included in the communications workshop?
    f. aspects of business communication
    g. advice on taking notes on business communications
    h. key concepts that will be discussed during the seminar
    j. useful strategies for organizing your work force
    k. how to help your team by making positive statements

25. The leaders of the company authorized this communications workshop. This fact suggests that the company leaders
    a. have recently identified a sharp decline in productivity.
    b. believe in the importance of communication among employees.
    c. think that employees should be seen and not heard.
    d. plan to sponsor communication workshops for other businesses.
    e. hope to increase the number of employees in management.

26. In addition to the Communication Workshop Overview, each participant was also given an organized chart in which to take notes on the conference. Which of the following conclusions about the leaders of the workshop is supported by this information and the excerpt?

    The workshop leaders

    f. believe that note taking is the most important communication skill of all.
    g. do not trust the participants to remember anything they hear.
    h. want to help employees develop the beneficial habit of note taking.
    j. think that note taking can take the place of verbal communication.
    k. will score participants on the quality and completeness of their notes.

Read the following passage, then answer Questions 27 through 30.

Kwan was anything but shy. She dropped her bag, fluttered her arms and bellowed, "Hall-oo! Hall-oo!" Still hooting and laughing, she jumped and squealed the way our new dog did whenever we let him out of the garage.

This total stranger tumbled into Mom's arms, then Daddy Bob's. She grabbed Kevin and Tommy by the shoulders and shook them. When she saw me, she grew quiet, squatted on the lobby floor and held out her arms. I tugged on my mother's skirt. "Is *that* my big sister?"

Mom said, "See, she has your father's same thick, black hair."

I still have the picture Aunt Betty took: curly-haired Mom in a mohair suit, flashing a quirky smile; our Italo-American stepfather, Bob, appearing stunned; Kevin and Tommy mugging in cowboy hats; a grinning Kwan with her hand on my shoulder; and me in a frothy party dress, my finger stuck in my bawling mouth.

I was crying because just moments before the photo was taken, Kwan had given me a present. It was a small cage of woven straw, which she pulled out of the wide sleeve of her coat and handed to me proudly. When I held it up to my eyes and peered between the webbing, I saw a six-legged monster, fresh-grass green, with saw-blade jaws, bulging eyes, and whips for eyebrows. I screamed and flung the cage away.

At home, in the bedroom we shared from then on, Kwan hung the cage with the grasshopper, now missing one leg. As soon as night fell, the grasshopper began to chirp as loudly as a bicycle bell warning people to get out of the road.

After that day, my life was never the same. To Mom, Kwan was a handy baby-sitter, willing, able, and free. Before my mother took off for an afternoon at the beauty parlor or a shopping trip with her gal pals, she'd tell me to stick to Kwan. "Be a good little sister and explain to her anything she doesn't understand. Promise?" So every day after school, Kwan would latch on to me and tag along wherever I went. By the first grade, I became an expert on public humiliation and shame. Kwan asked so many dumb questions that all the neighborhood kids thought she had come from Mars. She'd say: "What M&M?" "What ching gum?" "Who this Popeye Sailor Man?"

—"The Girl with Yin Eyes," from *The Hundred Secret Senses* by Amy Tan, copyright © 1995 by Amy Tan. Used by permission of G.P. Putnam's Sons, a division of Penguin Group (USA) Inc.

27. Which of the following best tells what this passage is about?
   a. The narrator meets her older sister for the first time.
   b. Kwan decides to run away but the narrator talks her out of it.
   c. Kwan brings the narrator a grasshopper in a cage as a gift, but the narrator is afraid of the creature.
   d. The narrator has to teach her sister who Popeye is.
   e. The narrator's life is changed when her older sister comes to live with her family.

28. Which of the following is the most likely reason that Kwan holds out her arms when she first meets the narrator?
   f. She wants the narrator to stay away from her.
   g. She wants to give the narrator a hug.
   h. She wants to stop herself from falling on the narrator.
   j. She wants to show that she has nothing in her hands.
   k. She wants the narrator to reach up her sleeve.

29. Kwan helps the family by
  a. running errands.
  b. translating.
  c. taking the narrator to school.
  d. babysitting the narrator.
  e. tutoring the narrator.

30. Judging from the passage, Kwan must be
  f. new to this country.
  g. a member of an unknown family.
  h. a distant relative.
  j. a college student.
  k. a neighbor.

Read the following passage, then answer Questions 31 through 35.

Over the last century, the average global temperature has risen about 1 degree Fahrenheit. In Alaska, the average temperature has increased by about 5 degrees just in the last thirty years. Alaska, as well as northern Canada and Russia, are clearly warming faster than Earth as a whole.

The warmer weather may provide some long-range benefits to Alaska's economy. A longer summer would boost tourism, and warmer water might increase production at the state's fisheries. Less sea ice would make drilling for offshore oil more profitable and the shipping of goods easier. It is entirely possible that an open sea route from Alaska to Europe will open up in the near future.

The warmer temperatures have also brought liabilities. One problem is the damage to Alaska's more than 1,000 glaciers. One such glacier, the Columbia, had been stable since 1923. It would pull back a bit every summer, then advance again in the winter and reattach itself to the broad, rocky, underwater shoal along the coast of Prince William Sound. In 1983, however, the Columbia Glacier failed to reattach. Since then, it has been retreating at an average rate of one-half mile each year. Most of the other glaciers in Alaska are retreating as well.

The warmer temperatures are also causing the permafrost—ground that was previously always frozen—to melt. Often, when permafrost thaws, a large hole, known as a thermokarst, opens up in the ground. Throughout Alaska, the number of thermokarsts is growing, and these holes cause damage to the infrastructure, such as buildings, bridges, and roads. In addition, huge patches of forests are drowning in thawing permafrost. The trees, destabilized by the melting layer beneath them, begin to tilt at crazy angles, resulting in a phenomenon called "drunken forest."

The forests in Alaska most affected by global warming are the boreal, or northern, forests. These forests are also known by the Russian word *taiga*. In addition to drowning in the melting permafrost, many of the trees have been weakened by the climate change and then killed by insects that only survive because of the warmer temperatures. Many scientists think the taiga will slowly move north, replacing the treeless tundra, if the warming continues.

Researchers also believe that Alaska's thaw is the result of both human-caused and natural changes. Although climate predictions are always uncertain, a trend seems to be developing. If temperatures continue to rise each year, scientists may be willing to say that Alaska is in the beginning stages of an entirely new climatic state.

31. Which of the following best tells what this passage is about?
  a. liabilities of warmer temperatures on Alaska's landscape
  b. effects of global warming in the Northern Hemisphere
  c. ever-shifting weather patterns of a new worldwide climatic state
  d. real and predicted effects of global warming in Alaska
  e. suggestion that the economy in Alaska is heating up

32. The passage suggests that some buildings in Alaska are threatened because
  f. frozen ground can crack a building's foundations.
  g. the ground on which the structures are built is unstable.
  h. the thermokarst holes prevent a lot of new construction.
  j. the harsh freezing erodes the walls of the buildings.
  k. the state supports tourism over structural maintenance.

33. According to the passage, what changes will occur in Alaska's tundra if temperatures continue to rise?
  a. Boreal forests will be replaced by the taiga.
  b. The shoal along the coast will melt.
  c. Trees will finally be able to grow there.
  d. Trees will drown in the melting permafrost.
  e. Oil drilling will increase.

34. The information in this passage suggests that, as compared to Alaska, northern Canada
  f. is experiencing the same warming effects.
  g. has more to gain from warming trends.
  h. does not have as many glaciers.
  j. does not have problems with thermokarsts.
  k. is experiencing fewer liabilities due to warming trends.

35. Based on the information in this passage, which of the following statements about the Columbia Glacier is accurate?
  a. It is Alaska's largest glacier.
  b. It has been stable since the early nineteenth century.
  c. It is a retreating land mass in Prince William Sound.
  d. Its underwater shoal reattaches to land each winter.
  e. It is a damaged glacier that is steadily retreating.

Read the following passage, then answer Questions 36 through 39.

Fighting fires has always been a dangerous job. The job has grown even more perilous in recent years. That's because changes in building construction over the last 30 years have resulted in hotter fires that are more unpredictable. Today's buildings use many more synthetic materials than did those built 30 years ago. As a result, when they catch fire, they burn at a much higher temperature and release much more energy than do their older counterparts. Furthermore, most modern buildings are extremely well insulated. That means that when they catch fire, they retain most of the fire's heat.

Hotter fires result in especially dangerous situations for firefighters. The fires themselves are hazardous, of course, but even more problematic are the super-hot gases created by the fires. These cause "extreme fire behaviors," a term firefighters use to describe quick-moving, life-threatening events at the scene of a fire. They include flashovers, temperature increases so great that all flammable materials in an area combust at once. They also include backdrafts, during which oxygen rushes into an area, causing an underfueled fire to burst into a fireball. One dangerous effect of a backdraft is that it can literally suck all the oxygen out of a room where firefighters are working. Finally, there are gas explosions. They occur when extreme heat causes atmospheric gases to expand quickly.

American firefighters have worked hard to combat the dangers of modern firefighting. Unfortunately, they have only been able to battle them to a standstill, and death rates have remained essentially unchanged during the last 30 years. That's a poor showing compared to Sweden and Great Britain. Both of those countries have cut death rates among firefighters in half over the same period.

Why are the British and the Swedes more successful at protecting their firefighters? Some within the firefighting community believe it's because both countries use a procedure called "3-D firefighting."

In 2002, the fire department of Gresham, Oregon (pop. 130,000) became the first in the United States to embrace 3-D firefighting techniques. The company's battalion chief, Ed Hartin, has become an impassioned advocate for the procedure, conducts trainings, and has coauthored a book on the subject.

36. Which of the following best tells what this passage is about?
   a. how the temperature increase in a flashover causes fireballs
   b. why firefighting is less dangerous in America now than in previous years
   c. The fire department of Gresham, Oregon was the first American company to use a new technique.
   d. Sweden and Great Britain have cut death rates among firefighters.
   e. the dangers of firefighting and the possibilities of a new technique

37. Which of the following conclusions about backdrafts does the passage support?
   f. Backdrafts cause all flammable materials in an area to combust at once.
   g. A backdraft can result in the suffocation of firefighters.
   h. A backdraft is easily stopped by releasing more oxygen into the room.
   j. Backdrafts occur when building insulation catches on fire.
   k. Backdrafts are always accompanied by gas explosions and flashovers.

38. The author supports the argument that the United States has made a "poor showing" in preventing firefighter deaths by
   a. pointing out that the death rate among U.S. firefighters has increased dramatically in the last 30 years.
   b. pointing out that the death rate among U.S. firefighters has decreased dramatically in spite of using old techniques.
   c. comparing U.S. death rates among firefighters with death rates in countries where 3-D firefighting is used.
   d. suggesting that U.S. fire departments are not really interested in reducing death rates among fire departments.
   e. suggesting that no U.S. fire departments are informed about 3-D firefighting techniques.

39. Information in the passage supports which of the following conclusions?
   f. Synthetic materials burn hotter than do natural materials.
   g. Gas explosions are the most dangerous form of extreme fire behavior.
   h. Backdrafts cannot be prevented with any known techniques.
   j. Extreme fire behaviors are uncommon today.
   k. Flashovers are the most dangerous form of extreme fire behavior.

Read the following passage, then answer Questions 40 through 45.

Firefighters in Sweden and Great Britain have had some success with a relatively new technique called 3-D firefighting. In the 3-D method, firefighters use thermal imaging devices to locate the hottest spots within a fire. These are the areas in which dangerous gases are most likely to accumulate, and thus are the sources of many extreme fire behaviors. Once they find the trouble spots, firefighters spray them with quick bursts of fine, cool mist. The mist lowers the temperature enough to allow firefighters to enter the area safely, identify the source of the fire, and quickly douse it with water. Proponents of the technique claim that it greatly reduces the likelihood of an extreme fire behavior.

Many American fire companies remain resistant to the 3-D method. Why? According to some, American firefighters are devoted to the traditions of their profession. They are equally suspicious of new technologies and techniques. Many intuitively believe that the 3-D technique cannot be effective. They wonder how such a small amount of water can have any noticeable effect on an intense fire. Many reject it simply because the technique contradicts one of the long-held truisms of firefighting: Never put water directly on smoke, because it could result in steam so hot that it will scald nearby firefighters.

Supporters of 3-D firefighting note that their technique uses too little water to create dangerous steam. They also argue that the amount of water used is ample to cool hot atmospheric gases. Finally, they point to the Swedish and British successes in reducing death rates as evidence that 3-D firefighting is both safe and effective. Criticism of 3-D firefighting, they argue, is based on old prejudices and false assumptions.

For now, most American fire companies will stick with traditional firefighting strategies. They will douse fires with a steady stream of water aimed directly at the source of the fire. Occasionally they will crisscross the ceiling with a stream of water in order to cool atmospheric gases (firefighters call this technique penciling). And they will adhere to their time-tested method of dealing with extreme fire behavior: They will duck and hope for the best.

Ultimately, the greatest obstacle to 3-D firefighting in the United States may be its cost. The technique requires the use of thermal imagers, which can cost upward of $10,000 apiece. For many fire departments, this expense may render any debate over the effectiveness of 3-D firefighting academic. After all, if a fire department can't afford a new technique, it probably isn't going to waste much time determining whether it is beneficial.

40. Which of the following best tells what this passage is about?
   a. why the best firefighting technique in the world is only being used in Europe
   b. the pros and cons of a new firefighting technique being used with some success in Europe
   c. how 3-D firefighting has revolutionized firefighting around the world
   d. 3-D firefighting techniques offer the most effective way to fight fires.
   e. why the 3-D firefighting method has been rejected in most countries

41. Which of the following is NOT one of the methods used in 3-D firefighting?
   f. quick bursts of fine, cool mist
   g. quickly dousing the fire with water
   h. using only a small amount of water
   j. dousing fires with a steady stream of water
   k. thermal imaging devices

42. The passage states that using a quick burst of fine, cool mist

    a. will probably result in the scalding of a firefighter.

    b. lowers the temperature enough to allow firefighters to enter the area.

    c. greatly increases the likelihood of an extreme fire behavior.

    d. has no noticeable effect on an intense fire.

    e. is only effective when used in "penciling."

43. Which of the following is NOT one of the reasons that many American companies resist using the 3-D method?

    f. Many intuitively believe that the 3-D technique cannot be effective.

    g. The technique contradicts one of the long-held truisms of firefighting: Never put water directly on smoke.

    h. They may be suspicious of new technologies and techniques.

    j. American firefighters are devoted to the traditions of their profession.

    k. The thermal imaging equipment used in the 3-D method is not reliable.

44. Opponents of 3-D firefighting would probably become more inclined to accept the 3-D technique if

    a. field tests proved that a small amount of water could significantly cool hot gases in the air.

    b. more countries used the 3-D firefighting techniques.

    c. the budgets of most American firefighting units were substantially reduced.

    d. no better method was developed.

    e. intense fires continued to claim the lives of American firefighters.

45. Which answer best describes the author's purpose in the last paragraph?

    a. to criticize U.S. firefighters for their unwillingness to try new techniques

    b. to present evidence that 3-D firefighting techniques are not effective

    c. to explain that 3-D firefighting techniques have applications outside the world of firefighting

    d. to convince readers that 3-D firefighting would reduce firefighting deaths in the United States

    e. to suggest one factor that might prevent wider use of 3-D firefighting techniques

Part 2—Math

The Math Test includes 50 questions covering content in the following areas:

- basic math
- percentages, fractions, decimals, averages
- pre-algebra
- algebra
- substitution
- factoring
- geometry
- probability
- logic
- word problems

Solve each problem and select the best answer from the choices given. It is important to keep in mind that:

- Formulas and definitions of mathematical terms and symbols are not provided.
- Diagrams other than graphs are not necessarily drawn to scale. Do not assume any relationship in a diagram unless it is specifically stated or can be figured out from the information given.

- A diagram is in one plane unless the problem specifically states that it is not.
- Graphs are drawn to scale. Unless stated otherwise, you can assume relationships according to appearance. For example, (on a graph) lines that appear to be parallel can be assumed to be parallel; likewise for concurrent lines, straight lines, collinear points, right angles, and so on.
- You will need to reduce all fractions to lowest terms.

46. Which of the following is an example of an irrational number?

a. $\sqrt{7}$

b. $\sqrt{9}$

c. $\sqrt{16x^2}$

d. $\sqrt[3]{-27}$

e. $\frac{3}{8}$

47. If the two horizontal lines in the figure below are parallel, what is the measure of $\angle B$ in the diagram below?

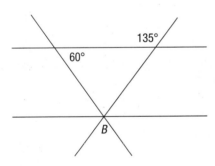

f. 45°

g. 60°

h. 75°

j. 55°

k. cannot be determined from the information given

48. Lefty keeps track of the length of each fish that he catches. Below are the lengths in inches of the fish that he caught one day:

$$12, 13, 8, 15, 10, 8, 9, 17$$

What is the median fish length that Lefty caught that day?

a. 8 inches

b. 11.5 inches

c. 11 inches

d. 17 inches

e. 12.5 inches

49. A merchant buys a product for $12.20 and then marks it up 35% to sell it. What is the selling price of the item?

f. $4.27

g. $7.93

h. $16.47

j. $20.13

k. $28.67

50. When graphed on the coordinate plane, what is the range of the function shown below when $-10 < x < 10$?

$$f(x) = \begin{cases} -x \text{ if } -10 < x < 0 \\ x^2 \text{ if } 0 \le x < 10 \end{cases}$$

a. all real numbers y such that $-10 < y < 10$

b. all real numbers y such that $10 < y < 100$

c. all real numbers y such that $0 < y < 100$

d. all real numbers y such that $0 < y < 100$

e. all real numbers y such that $0 \le y < 100$

51. $\left(\frac{2x}{3y} \div \frac{4x^2}{9y} \right) \div \left(\frac{7xy^2}{10x^3} \right)$

f. $\frac{56x}{270}$

g. $\frac{80x^5}{189y^4}$

h. $\frac{15x}{7y^2}$

j. $\frac{90xy^2}{42}$

k. $\frac{180x^7}{84y^4}$

52. The sum of the interior angles of a polygon is 18y. If y is 15 greater than the number of sides of the polygon, how many sides does the polygon have?

a. 5
b. 6
c. 7
d. 10
e. 13

53. In the graph, what is the overall ratio of males to females for the three rooms?

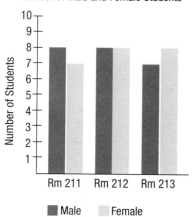

Number of Male and Female Students

f. 7:1
g. 1:7
h. 1:1
j. 7:8
k. 8:7

54. If the average of five consecutive even whole numbers is 48, what is the largest number?

a. 45
b. 46
c. 49
d. 52
e. 54

55. Which of the following is equivalent to $a^1a^{-2}a^3b^{-1}b^2b^{-3}$?

f. 1
g. ab
h. a^2b^2
j. $\frac{a}{b}$
k. $\frac{a^2}{b^2}$

56. Tyson and Steve both collect skateboards. Tyson owns three less than seven times the number of skateboards Steve owns. If s represents the number of skateboards Steve owns, which of the following expressions represents the number of skateboards Tyson owns?

a. $7s$
b. $3 - 7s$
c. $7s - 3$
d. $(7s)(3)$
e. $(\frac{1}{7})s + 3$

57. If an angle in a rhombus measures 21°, then the other three angles consecutively measure

f. 159°, 21°, 159°.
g. 21°, 159°, 159°.
h. 69°, 21°, 69°.
j. 21°, 69°, 69°.
k. cannot be determined from the information given

58. Mark, Todd, Rachel, John, and Mary's names were each written on a piece of paper and tossed into an empty bag. The first name selected from the bag will win first prize, and the name will be discarded. Next, a second-prize winner will be selected from the remaining names. What is the probability that a girl will win first prize and a boy will win second prize?

a. $\frac{5}{5}$
b. $\frac{3}{10}$
c. $\frac{6}{25}$
d. $\frac{7}{20}$
e. $\frac{6}{9}$

59. 5.133 multiplied by 10^{-6} is equal to

   f. 5,133,000

   g. 0.0005133

   h. 0.00005133

   j. 0.000005133

   k. 0.0000005133

60. The number of gallons of paint needed to cover the exterior of a house with one coat of paint is estimated by the formula $[10n(l + w) - 9n] \div 350$, where n is the number of stories, l is the length of the house in feet, and w is the width of the house in feet. Approximately how many gallon cans of paint should somebody buy in order to paint one coat on the exterior of a 30×50-foot 2-story house?

   a. 3

   b. 4

   c. 5

   d. 6

   e. 7

61. For Event A, the probability of Event A occurring may be all of the following EXCEPT:

   f. 0.56

   g. 1.5

   h. $\frac{1}{2}$

   j. 0.3333333 . . .

   k. 1

62. If q is a prime number that is less than 10, what is the largest possible value of $(2q) + 1$?

   a. 19

   b. 15

   c. 10

   d. 7

   e. 8

63. Quadrilaterals $ABCD$ and $EFGH$ are similar. If the perimeter of quadrilateral $ABCD$ is equal to $4y^2$, what is the perimeter of quadrilateral $EFGH$?

   f. y^2

   g. $2y$

   h. $4y^2$

   j. $16y^4$

   k. cannot be determined from the information given

64. In Mrs. Marsh's class, 3 out of every 7 students are boys. If there are 28 students in her class, how many boys are in her class?

   a. 3 boys

   b. 7 boys

   c. 12 boys

   d. 14 boys

   e. 16 boys

65. If $w@z$ is equivalent to $3w - z$, what is the value of $(w@z)@z$?

   f. $3w - 2z$

   g. $6w - 2z$

   h. $9w - 4z$

   j. $9w - 3z$

   k. $9w - 2z$

66. Lara is in charge of ticket sales for the school play. A student ticket costs $4.50 and an adult ticket cost $8.50. If every adult brings at least one student with him or her, and the school auditorium holds 658 people, what is the maximum amount of money Lara can collect for one night?

   a. $5,593.00

   b. $4,277.00

   c. $2,961.00

   d. $2,632.00

   e. $4,056.00

67. In the diagram, line *AB* is parallel to line *CD*. Transversal *MN* intersects *AB* and *CD* at *P* and *Q* respectively. If $m\angle APQ = 2x + 20$ and $m\angle PQC = 3x - 10$, what is $m\angle DQN$?

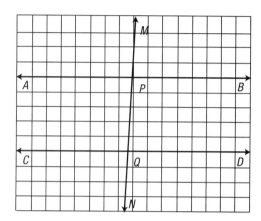

   f. 34°
   g. 88°
   h. 92°
   j. 98°
   k. 146°

68. The price of a $220 saw was reduced to $154. What was the percent discount?
   a. 30%
   b. 43%
   c. 66%
   d. 67%
   e. 70%

69. At a flower shop, there are 49 red plants and 15 seedless plants. Of these, 14 plants are both red and seedless. There are also 5 plants that are neither seedless nor red. Matt runs into the shop and randomly purchases a plant. What is the probability that he purchased a seedless red plant?
   f. $\frac{3}{11}$
   g. $\frac{14}{69}$
   h. $\frac{15}{83}$
   j. $\frac{14}{55}$
   k. $\frac{14}{83}$

70. What is the perimeter of an isosceles right triangle whose hypotenuse is $5\sqrt{2}$ units?
   a. 5 units
   b. $15 + \sqrt{2}$ units
   c. $10 + 5\sqrt{2}$ units
   d. $15\sqrt{2}$ units
   e. $25\sqrt{2}$ units

71. Pat has been contracted to replace a countertop in a kitchen. Pat will be paid $1,000 to do the job, that can be completed in 3 days working 8 hours a day. She will need to pay the cost of the materials, which is $525 out of her $1,000 pay. How much will Pat's hourly pay be for the job?
   f. $59.38
   g. $21.89
   h. $65.63
   j. $23.56
   k. $19.79

72. If $a = -3$, $b = -4$, and $c = -5$, which of the following has the least value?
   a. $|a| + |b|$
   b. $|c| \times |b|$
   c. $a \times |b + c|$
   d. $a \times |b|$
   e. $b \times c$

73. If both 3 and 5 are factors of the odd number *Q*, what might the value of *Q* be?
   f. 36
   g. 70
   h. 65
   j. 45
   k. 30

74. What is the area of the shaded region in the figure below?

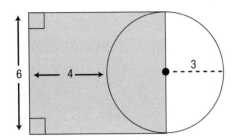

a. $24 - 9\pi$ square units
b. $24 + 9\pi$ square units
c. $24 - 6\pi$ square units
d. $24 - 4.5\pi$ square units
e. $42 - 4.5\pi$ square units

75. Jack is retiling his kitchen floor. Each tile is $1\frac{1}{8}$ foot by $1\frac{3}{5}$ foot. What is the area of each tile?

f. $\frac{3}{40}$
g. $\frac{3}{8}$
h. $1\frac{4}{5}$
j. $1\frac{3}{5}$
k. $1\frac{2}{5}$

76. Matt only likes shapes that have at least one pair of congruent sides. Matt likes all of the following EXCEPT

a. squares.
b. trapezoids.
c. rhombuses.
d. parallelograms.
e. rectangles.

77. $\triangle ABC$ has coordinates $A(1,3)$; $B(4,5)$; $C(5,0)$. What are the coordinates of point C after a reflection over the y-axis?

f. $(5,0)$
g. $(0,5)$
h. $(-5,0)$
j. $(0,-5)$
k. $(5,-5)$

78. What is the slope of the line determined by the equation $15y - 10x = -9$?

a. $\frac{3}{2}$
b. $\frac{2}{3}$
c. $\frac{3}{5}$
d. $-\frac{2}{3}$
e. $-\frac{3}{2}$

79.

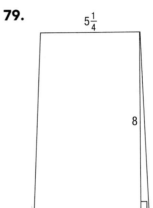

The formula for the area of a trapezoid is $A = \frac{1}{2}h(b_1 + b_2)$. If the area of the trapezoid is 45, which expression below gives the length of the other base?

f. $\dfrac{2(45)}{5\frac{1}{4}} - 8$

g. $\dfrac{2(45)}{8} - 5\frac{1}{4}$

h. $\dfrac{2(45)}{5\frac{1}{4}} - \dfrac{2(45)}{8}$

j. $\dfrac{(5\frac{1}{4})(45)}{8} - \dfrac{2(45)}{8}$

k. $\dfrac{(5\frac{1}{4})(45)}{8}$

80. If the height of a triangle is half its base, b, what is the area of the triangle?

a. $\left(\frac{1}{4}\right)b$
b. $\left(\frac{1}{4}\right)b^2$
c. $\left(\frac{1}{2}\right)b$
d. $\left(\frac{1}{2}\right)b^2$
e. b

81. Considering "If you are the president of the United States, then you must be over 35," which of the following is true?

   f. Being the president of the United States is necessary for being over 35.

   g. There is no age requirement to be president of the United States.

   h. Being 35 is necessary for being the president of the United States.

   j. It is sufficient to know that you are over 35 for the statement "you are the president of the United States" to be considered true.

   k. It is sufficient to know that if you are the president of the United States you may be under 35 years old.

82. If a and b are odd numbers and c and d are even numbers, which of the following equations would never hold true?

   a. $a \times c = d$

   b. $a + b = d$

   c. $c \times d = a$

   d. $d \div c = a$

   e. $d \div a = c$

83. Based on the diagram, if lines L and M are parallel, which of the following equations is NOT necessarily true?

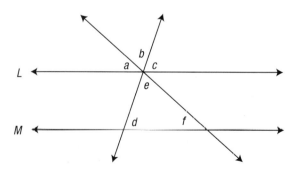

   f. $a + b + c = d + e + f$

   g. $a + c = 180 - e$

   h. $a + e + c = 180$

   j. $b + c = e + f$

   k. $a + b + c + d + e + f = 360$

84. If $|2x - 3| \geq 20$, which shows the solutions for x?

   a. $-\frac{17}{2} = x = \frac{23}{2}$

   b. $x \leq -\frac{17}{2}$ or $x \geq \frac{23}{2}$

   c. $x = \frac{23}{2}$

   d. $x \leq \frac{17}{2}$

   e. There is no solution.

85. The radius of a cylinder is $2x$ and the height of the cylinder is $8x + 2$. What is the volume of the cylinder in terms of x?

   f. $(16x^2 + 4x)\pi$

   g. $(16x^3 + 4x^2)\pi$

   h. $(32x^2 + 8x)\pi$

   j. $(32x^3 + 8x^2)\pi$

   k. $(128x^3 + 64x^2 + 8x)\pi$

86. Given $U = \{w \mid w$ is an integer$\}$, $A = \{x \mid x$ is a positive whole number and is less than 20$\}$, and $B = \{y \mid y$ is a negative odd integer and is greater than $-30\}$, which of the following statements is true?

   a. U is a subset of B.

   b. B is a proper subset of A.

   c. The cardinal number of set A is 20.

   d. Sets A and B are disjoint.

   e. $B' = A$

87. Mischa draws two isosceles right triangles. Daryl looks at the triangles, and without measuring them, he can be sure that the triangles are

   f. corresponding.

   g. similar.

   h. similar and congruent.

   j. congruent, but not similar

   k. neither similar nor congruent.

88. Main Street and John Street are parallel in the diagram below. Jackson Street intersects both streets and forms a 48° angle with Main Street as shown in the figure. What is the measure of ∠y?

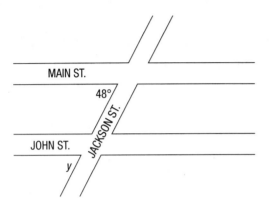

a. 132°

b. 42°

c. 48°

d. 90°

e. 32°

89. If the circumference of a circle measures 10 feet, what is its area in square feet?

f. 5π

g. $\frac{10}{\pi}$

h. 15π

j. $\frac{25}{\pi}$

k. 30π

90. Based on the number line below, which of the following inequalities is correct?

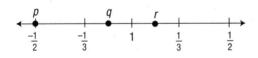

a. $|q| > p > |r|$

b. $q < |r| < |p|$

c. $p > q > r$

d. $|q| < p < r$

e. $|q| < p < |r|$

91. The slope of a line parallel to $y - 4 = 5(x + 7)$ is:

f. $-\frac{1}{5}$

g. −1

h. 1

j. $\frac{1}{5}$

k. 5

Use the diagram to answer questions 92 and 93:

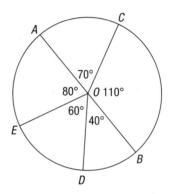

92. If the radius of the circle is 12 units, which sector has an area of 24π square units?

a. sector *EOD*

b. sector *DOB*

c. sector *BOC*

d. sector *AOC*

e. sector *EOA*

93. If the radius of the circle is 15 units, what is the area of sector *DOB*?

f. $\left(\frac{30}{9}\right)\pi$ square units

g. 22.5π square units

h. 25π square units

j. 45π square units

k. 225π square units

94. If *p* is divisible by both 20 and 4, which of the following statements is NOT true?

    a. *p* must be divisible by 5.

    b. *p* must be divisible by 1.

    c. *p* must be divisible by 2.

    d. *p* must be divisible by 10.

    e. *p* must be divisible by 80.

95. In the diagram, what is the value of *x*?

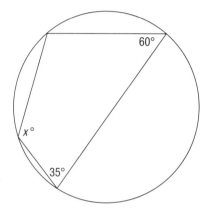

    f. 55

    g. 60

    h. 90

    j. 120

    k. cannot be determined

Answers

Paragraph 1 (Q, T, S, U, R) Only sentence **Q** logically follows the given introductory sentence. Sentence **R** obviously comes last, since it describes what happens later than the main part of the paragraph. The best order for the remaining sentences, all of which refer to the leaves described in **Q**, is **T, S, U**. The insects are attracted (**T**), then they land (**S**), and then the leaves snap shut (**U**).

Paragraph 2 (S, Q, U, R, T) Sentence **S** refers back to the beach agates introduced in the given sentence. Sentence **Q** tells how the agates were formed, referring back to sentence **S**. Sentence **U** refers back to the oxides and minerals mentioned in sentence **Q**. Sentence **R** refers back to the colors mentioned in sentence **U**. Sentence **T** refers back to the colors and polishing mentioned in sentence **R**.

Paragraph 3 (R, U, S, Q, T) At first, you might not be certain which sentence should follow the given introductory sentence. Pay close attention to the repetition of words from sentence to sentence, and the choices will fall into their correct order. Sentence **U** refers back to the transformation mentioned in sentence **R**. Sentence **U** introduces living creatures, and sentence **S** is about those creatures that survive. Sentence **Q** is about the surviving creatures passing on favorable changes to their offspring. So these sentences logically come in the order **U, S, Q**. Sentence **T** gives Darwin's name for the process explained in the earlier sentences.

Paragraph 4 (U, R, T, Q, S) Sentence **U** follows up on the mention of photographs from the introductory sentence and introduces the idea of using both a photograph and a sketch. Sentence **R** says that the sketch may be done before the photograph and sentence **T** says that "then the camera" records details and fills in "whatever the sketch may have missed," indicating that **T** follows **R**. In sentence **Q**, the words "camera also records" indicate that **Q** comes after **T**. Sentence **S** sums up the information already given and is best used as this paragraph's concluding sentence.

Paragraph 5 (Q, T, S, U, R) Sentence **Q** describes the problem of not listening well. Sentence **T** states that it is the first step toward improvement, and sentence **S** is the second step. Sentence **U** gives examples of the active response mentioned in sentence **S**. In sentence **R**, the words "also respond actively" indicates that it comes after another mention of an active response, so **R** follows **U**.

6. c. Based only on the information given, we do not know whether Christopher is usually forgetful (choice **a**) or how many friends he has (choice **b**). We can conclude that Christopher will probably not draw tonight (choice **c**), since we are told that for his drawings, he uses his special pen only, which he has forgotten at a friend's apartment. It is possible that Christopher draws every night (choice **d**) or decided to use a different pen (choice **e**), but those statements are unsupported by the facts given.

7. h. Sexton is farther away than Fromme, who is five miles away, and closer than Smith, who is seven miles away. Therefore, Sexton must be six miles away.

8. d. Based on the statements given, we can conclude that yesterday was a cold and rainy day (choice **d**), since we are told that Dan spent an entire day at home, and we know that Dan never stays in when the weather is nice. We are not told what Dan does when he stays at home (choices **a** and **b**). We can also assume that if the weather was nice yesterday (choice **c**), Dan would not have stayed in.

9. k. The first two statements being true, both Zoe and Heather read more books than Jenna, but it doesn't tell us whether Zoe read more than Heather.

10. b. The best way to answer this question is to create a map like the one below.

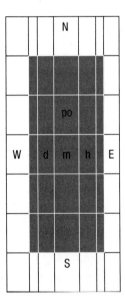

11. k. Based on the available information, the only thing we can conclude for certain is that it takes approximately six hours by plane to get to Vancouver, BC, from New York City. We do not know how soon Jenny will arrive to Vancouver, BC (choice **g**), because we do not know how long she has been on the plane. Choices **f**, **h**, and **j** may be true, but we do not have enough evidence to support them.

12. e. If Georgia is older than Marsha and Bart is older than Georgia, then Marsha has to be the youngest of the three. Choice **b** is clearly wrong because Bart is the oldest. There is no information in the paragraph to support either choice **a** or choice **c**. Marsha wins more often than Georgia, but how often Bart wins is not mentioned, so we cannot determine whether choice **d** is true.

13. j. The only thing we can conclude for certain is that Linda is not one of the Petersens' children, since we are told that the Petersens have three children: Pete, Jim, and Michele. Linda cannot be younger than Michele (choice **f**) because we are told that Linda is 6 and Michele is 5. There is nothing to suggest that Linda is a daughter of the Petersens' neighbors (choice **h**) or a cousin (choice **k**). Linda may certainly be Michele's friend (choice **g**), but we do not have enough evidence to support this option.

14. d. Based only on the information given, the only correct assumption we can make is that Fellini was born and raised in Italy, because we are told that just like Fellini, Mr. Magnani was born and raised in Italy. Choice **a** may be true, but there is nothing in the statements given to us that supports this conclusion. Choices **b** and **c** may also be true, but there is no way to verify this information. Nothing indicates that the two men ever met (choice **e**).

15. j. We can tackle this question is with a simple diagram. The original schedules for Ms. Johnson, Mr. Carter, and Mr. Kirk are given. Since we know who traded days, we can fill in the final schedule.

	Original	After Switch
M	Ms. Johnson	Mr. Carter
T		Mr. Kirk
W	Mr. Carter	Ms. Johnson
Th	Mr. Kirk	Mr. Falk

After all the switches were made, Mr. Kirk worked on Tuesday. Mr. Carter worked on Monday, Ms. Johnson on Wednesday, and Ms. Falk on Thursday.

16. c. This is stated in the first two sentences, then developed in the rest of the passage. The remaining choices refer to details within the passage.

17. g. This is stated explicitly in the second sentence of the passage. Choice **f** is incorrect because only bad *eris* was defined as violent. Choice **h** deals with problems that belong in the domain of mankind, not the universe. Choices **j** and **k** have no support in the passage.

18. a. This is a definition explicitly stated in the sixth sentence. Choice **b** is incorrect because a choice dealing with mankind alone is too narrow for a definition of *eris*, which deals with the entire universe. Choice **c** is incorrect because it only deals with one action of the personified concept in goddess form. Choices **d** and **e** have no support in the passage.

19. j. This is stated in the third sentence of the second paragraph. Zeus did not want to sire a child who could eventually overthrow him. According to the passage, he felt it was safer to arrange for the child's father to be part mortal. There is no support in the passage for any of the other choices.

20. c. This answer follows the logic of the previous answer. That a mortal child "could never challenge the gods" implies that Zeus feared that if the child were immortal, it would overthrow him. The other choices mention individual words that appear in the passage, but are not given as a reason for Zeus not fearing the child of King Peleus.

21. g. The passage tells us that Achilles was the son of Thetis and Peleus, and the last paragraph states that the war will result in Achilles's death. Choice **f** is incorrect because there is no other mention of Zeus or events in the Trojan War other than Achilles's death. Choice **h** is incorrect because Eris purposely created the conditions that would lead to the war to kill the child of the bride and groom. Choices **j** and **k** are incorrect because Achilles is the son of Thetis and Peleus, the bride and groom of the myth.

22. e. This lesson is discussed explicitly in the last paragraph. All other choices are irrelevant.

23. e. The overall subject of the conference is improving your business communication skills. The other choices are either inaccurate (**a** and **d**) or define the topic too narrowly (choices **b** and **c**).

24. j. Note that the question asks for the item not included in the communications workshop. Although the overview refers to the work of a team, it does not say anything about organizing a work force. The other choices are included in the workshop.

25. b. The fact that the leaders of the company authorized this communications workshop suggests that they believe in the importance of employee communication. The other choices are not supported by the text.

26. h. The fact that the leaders provide a note-taking chart, combined with the support for effective note taking in the overview, suggests that the leaders want to help employees learn how to take effective notes. The other conclusions are not supported by the text.

27. e. Even though this is only stated clearly in the final paragraph, this is the only choice broad enough to cover the entire passage. Choices **a**, **c**, and **d** are true, but are about details in the passage. Choice **b** is not included in the passage.

28. g. Kwan immediately says hello to the other family members by hugging or touching them. The passage says that Kwan squats on the floor before holding out her arms to the narrator. She is not described as falling (choice **h**). Choice **j** makes no sense. Choice **k** is incorrect because the excerpt says that Kwan later reaches up her own sleeve to pull out a surprise for the narrator.

29. d. The passage says that the narrator's mother used Kwan as a babysitter. The other choices are not supported by the passage.

30. f. The narrator has not met Kwan before and Kwan clearly knows little about the United States. The passage identifies Kwan as the narrator's sister, therefore **g**, **h**, and **k** are incorrect. The passage does not mention college (choice **j**).

31. d. This choice best summarizes what the passage is about. Choice **a** is too narrow to be a main idea because only a portion of the passage tells the benefits of warmer weather. Choice **b** is too broad, since the passage is only about Alaska. Choice **c** is unsupported by the text. Choice **e** is too narrow since only a portion of the passage deals with this topic.

32. g. This is an inference based on the information in the fourth paragraph which states that the permafrost thaws, leaving holes that cause damage to infrastructure, including buildings. Choice **f** is incorrect because the passage indicates that permafrost in its always frozen state is not a problem. It's the thawing of the permafrost that causes the problem. Choice **h**—though possible—is unsupported by the information in the passage. Choice **j** is incorrect because it is not the freezing that is the problem; it's the thawing. Choice **k** is not supported by the information in the passage.

33. c. This detail is supported by the information in the closing sentence of the fifth paragraph—"Many scientists think the taiga [forests] will slowly move north, replacing the treeless tundra. . . ." Choice **a** makes no sense because the boreal forests are the same as the taiga. Choice **b** is similarly nonsensical because a shoal is rocky and doesn't melt. Choice **d** refers to what is happening to trees in Alaska's forests, not its treeless tundra. There is no evidence to support choice **e**.

34. f. According to the first paragraph, Alaska, northern Canada, and Russia are clearly warming faster than Earth as a whole. The other choices are all unsupported by the text.

35. e. This is the only fully accurate statement based on the information in the third paragraph. Since 1983, the glacier's ability to advance and reattach each year has been hindered due to warmer temperatures. The paragraph states that since 1983, the glacier has retreated an average of one-half mile per year. Choice **a** is unsupported by the passage. Choice **b** is incorrect because the glacier has been stable since 1923—the early twentieth century, not the nineteenth. Choice **c** is incorrect because the glacier is not a land mass; it is ice, and therefore is being affected by the warmer temperatures. Choice **d** describes the behavior of the glacier prior to 1983 but is no longer an accurate statement about the glacier's behavior.

36. e. Only the last choice is broad enough to tell what the passage is about. Choice **a** is wrong because the passage states that backdrafts, not flashovers, cause fireballs. Choice **b** is contradicted by the passage. The remaining choices are only details from the passage.

37. g. The passage states that "One dangerous effect of a backdraft is that it can literally suck all the oxygen out of a room where firefighters are working." When there is no oxygen in a room, people in that room are in danger of suffocating; therefore, the passage supports the conclusion stated in choice **g**. The passage never discusses the effects of burning building insulation, so choice **j** cannot be the best answer choice. The passage also mentions flashovers and gas explosions, but it never implies that they always occur when there is a backdraft (choice **k**). Choice **h** is not mentioned in the passage.

38. c. According to the third paragraph, "death rates have remained essentially unchanged during the last 30 years" in the United States. So these death rates have not increased (choice **a**) or decreased (choice **b**). Choice **c** is correct because the author compares U.S. death rates to those of Sweden and Great Britain, which have "cut death rates among firefighters in half over the same period." Choice **e** cannot be correct because the passage mentions the use of the 3-D technique in Oregon, and a book on the subject. There is no information in the passage to support choice **d**.

39. f. The first paragraph of the passage states that "Today's buildings use many more synthetic materials than did those built 30 years ago. As a result, when they catch fire, they burn at a much higher temperature and release much more energy than do their older counterparts." Because nonsynthetic materials are by definition natural materials, the passage supports the conclusion that synthetic materials burn hotter than do natural materials. Choices **g** and **k** are incorrect. Although the passage discusses three types of extreme fire behavior, it does not compare the relative dangers of these three. Choice **j** is contradicted by the passage, which states that firefighters face hotter fires today and that hotter fires cause extreme fire behaviors. Choice **h** is not covered in the passage.

40. b. The passage describes the firefighting technique and includes the opinions of supporters and of those who are skeptical about it. According to the passage, the effectiveness of 3-D firefighting is a controversial subject. Because there is strong disagreement on the subject, the passage does not support choice **d**. The remaining choices are not supported by the passage.

41. j. Note that the question asks for the exception. Dousing fires with a steady stream of water is described as a traditional technique, not as a 3-D firefighting technique. The other choices are described as part of the 3-D technique.

42. b. The first paragraph states that the mist of water "lowers the temperature enough to allow firefighters to enter the area safely, identify the source of the fire, and quickly douse it with water." Choice **c** is contradicted in the first paragraph. Choices **a** and **d** are noted as concerns about the method, but nothing indicates that these actually happen. Choice **e** is a traditional firefighting technique that uses a stream of water, not a fine mist.

43. k. Note that the question asks which choice is not a reason for rejecting the 3-D method. There is no indication in the passage that the thermal imaging equipment is not reliable. The other choices *are* given as possible reasons why American companies resist the method.

44. a. In the second paragraph, the passage states that opponents of 3-D firefighting "wonder how such a small amount of water can have any noticeable effect on an intense fire." Therefore, one of the reasons that they are skeptical about the technique is that they don't believe that a fine mist could cool such a hot fire. If field tests proved otherwise, opponents would probably be more likely to accept the technique. Lower budgets (choice **c**) would not help because the expense is one reason that U.S. firefighters don't use the 3-D method. The other choices are not given as reasons to oppose or accept the 3-D method.

45. e. The last paragraph discusses the high cost of 3-D firefighting equipment, noting that "this expense may render any debate over the effectiveness of 3-D firefighting academic. After all, if a fire department can't afford a new technique, it probably isn't going to waste much time determining whether it is beneficial." Thus, the author's purpose is to explain how the cost of the equipment might prevent wider use of the 3-D firefighting techniques. The other choices are not mentioned in the paragraph.

46. a. $\sqrt{7}$. All of the other choices are terminating decimals. $\sqrt[3]{-27}$ is imaginary, but it is rational.

47. h. 75°. Using corresponding angles and vertical angles, subtract 60 degrees from 135 degrees to isolate angle B.

48. c. **11 inches.** The median is the central number when the numbers are listed in increasing order and in this case the central two numbers are averaged to find the median: 8, 8, 9, **10, 12**, 13, 15, 17.

49. h. **$16.47.** Multiply the price by the markup percentage to get the markup amount: ($12.20)(0.35) = $4.27. Then add that to $12.20 to get the selling price.

50. e. **all real numbers y such that $0 \leq y < 100$.** The first requirement states that when x is between −10 and 0, the ranges will be between 0 and +10. The second requirement states that the range will go from 0 up to, but not including, 100, when x is between 0 and 10.

51. h. $\frac{15x}{7y^2}$.

$$\left(\frac{2x}{3y} \div \frac{4x^2}{9y}\right) \div \left(\frac{7xy^2}{10x^3}\right)$$
$$= \left(\frac{2x}{3y} \times \frac{9y}{4x^2}\right) \times \left(\frac{10x^3}{7xy^2}\right)$$
$$= \frac{180x^4y}{84x^3y^3}$$
$$= \frac{180x^4y}{84x^3y^3} = \frac{15x}{7y^2}$$

52. a. **5.** The sum of the interior angles of a polygon with n sides is $180(n-2)$, so $180(n-2) = 18y$. Since $y = 15 + n$, set up the equation, $180(n-2) = 18(15 + n)$. Then $10(n-2) = 15 + n$ and $n = 5$.

53. h. **1:1.** The total number of males is equal to the total number of females. 23:23 = 1:1.

54. d. **52.** Even numbers are always represented by "$2n$." Let the 5 even numbers be represented by $2n$, $2n + 2$, $2n + 4$, $2n + 6$, and $2n + 8$. The equation for their average will be $\frac{10n + 20}{5} = 48$. Multiplying both sides by 5 gives $10n + 20 = 240$, so $n = 22$. The largest number, $2n + 8$, will be 52.

55. k. $\frac{a^2}{b^2}$. Adding the exponents of the like bases gives a^2b^{-2} which is equivalent to $\frac{a^2}{b^2}$.

56. c. $7s - 3$. "Less than" means subtraction, but order of the terms is switched when using less than, so "−3" comes after $7s$.

57. f. **159°, 21°, 159°.** The adjacent angles in a rhombus are supplementary. The next angle is 159°, then 21°, then 159° again.

58. b. $\frac{3}{10}$. The probability of two events happening consecutively is the product of the individual probabilities. The probability that the first name selected from the bag will be a girl is $\frac{2}{5}$ since there are two girls and 5 people all together. Then there would be 3 boys out of 4 people total, so the probability that the second prize would go to the boy is $\frac{3}{4}$. The product of $\left(\frac{2}{5}\right) \times \left(\frac{3}{4}\right)$ is $\frac{6}{20} = \frac{3}{10}$.

59. j. **0.000005133.** Move the decimal 6 places to the left to get 0.000005133.

60. c. **5.** Since $n = 2$, $l = 30$, and $w = 50$: $[10 \times 2(30 + 50) - 9 \times 2] \div 350 = [20(80) - 18] \div 350 = 4.52$, so 5 gallons will be needed.

61. g. **1.5.** It is never possible to have a probability greater than 1.

62. b. **15.** The largest possible value of q is 7 so the largest value for $(2q) + 1 = 2 \times 7 + 1 = 15$.

63. k. **cannot be determined.** Since you do not have any information about the corresponding sides of these quadrilaterals, you cannot know what the ratio is between their sizes or perimeters.

64. c. **12 boys.** $\frac{boys}{total} = \frac{3}{7} = \frac{boys}{28}$. So, there are 12 boys in her class since the multiple between these two ratios is 4.

65. h. $9w - 4z$. Since $(w@z)@z = (3w - z)@z$, you will need to substitute $(3w - z)$ for w in the expression $3w - z$: $3(3w - z) - z = 9w - 3z - z = 9w - 4z$.

66. b. **\$4,277.00.** Since adult tickets are more expensive than student tickets, the greatest amount of money can be collected if the largest number of adults possible attend. Therefore if every adult brings only one student, the play will net the largest amount of money. Since there are 658 seats, if 329 adults and 329 students attend, Lara will collect $329(\$4.50) + 329(\$8.50) = \$4,277$.

67. h. **92°.** $\angle APQ$ and $\angle PQC$ are same-side interior angles, which means they are supplementary. Therefore, $(2x + 20) + (3x - 10) = 5x + 10 = 180$, so $x = 34°$. $\angle DQN$ is a vertical angle with $\angle PQC$ so they will have congruent measures. $m\angle DQN = m\angle PQC = 3(34) - 10 = 92$.

68. a. **30%.** Percent discount $= \frac{\text{difference in price}}{\text{original price}} = \frac{220 - 154}{220} = 0.3 = 30\%$.

69. j. $\frac{14}{55}$. The total number of plants in the shop is $49 + 15 - 14 + 5 = 55$. The number of seedless red plants is 14. So the probably that Matt bought a seedless red plant is $\frac{14}{55}$.

70. c. $10 + 5\sqrt{2}$ **units.** An isosceles right triangle is a 45-45-90 triangle with hypotenuse $=$ (leg)$(\sqrt{2})$. Therefore since the hypotenuse is $5\sqrt{2}$ units, the legs must each be 5 units long and the perimeter will be $10 + 5\sqrt{2}$ units.

71. k. **\$19.79.** After paying for materials, Pat will have $\$1000 - \$525 = \$475$. Pat will work for 24 hours over the three days, so Pat's hourly earnings will be $\frac{\$475}{24} = \$19.79$.

72. c. $a \times |b + c|$. The absolute value bars will turn any number inside positive:
$|a| + |b| = 7$
$|c| \times |b| = 20$
$a \times |b + c| = -27$
$a \times |b| = -12$
$b \times c = 20$

73. j. **45.** 45 is the only odd number listed that is divisible by both 3 and 5.

74. e. $42 - 4.5\pi$ **square units.** Find the area of the rectangle and subtract one-half the area of the circle in order to find the area of the shaded region.
Area of Rectangle $= 42$ square units (6×7)
Area of Circle $= 9\pi \times \frac{1}{2} = 4.5\pi$
Area of Shaded Region $= 42 - 4.5\pi$ square units

75. h. $1\frac{4}{5}$. Change each fraction into an improper fraction and then multiply them in order to find the area.
$1\frac{1}{8} = \frac{9}{8}$
$1\frac{3}{5} = \frac{8}{5}$
$\frac{9}{8} \times \frac{8}{5} = \frac{9}{5} = 1\frac{4}{5}$

76. c. **rhombuses.** A rhombus can have parallel sides, but does not *have* to have parallel sides. The rest of the shapes listed *must* have at least one pair of congruent sides by definition.

77. h. **(−5,0).** When reflecting a point over the y-axis, only its x-coordinate changes by switching signs.

78. b. $\frac{2}{3}$.
$15y - 10x = -9$ can be written in slope-intercept formula as follows:
$15y - 10x + 10x = 10x + -9$
$\frac{15y}{15} = \frac{10x + -9}{15}$
$y = \frac{2x}{3} - \frac{3}{5}$, where the slope is $\frac{2}{3}$.

79. g. $\frac{2(45)}{8} - 5\frac{1}{4}$.

Use the given formula and information, and then get b_2 by itself:

$45 = \frac{1}{2}(8)(5\frac{1}{4} + b_2)$

$45(2) = 8(5\frac{1}{4} + b_2)$

$\frac{45(2)}{8} = 5\frac{1}{4} + b_2$

$b_2 = \frac{45(2)}{8} - 5\frac{1}{4}$

80. b. $(\frac{1}{4})b^2$. Let the height of the triangle = $(\frac{1}{2}base)$ and substitute that into the formula for area: $A = \frac{1}{2}(base)(height) = \frac{1}{2}(b)(\frac{1}{2}b) = \frac{1}{4}b^2$.

81. h. Being 35 is necessary for being the president of the United States.

82. c. $c \times d = a$. The product of two even numbers can never be an odd number.

83. j. $b + c = e + f$. Angles b and e are equal since they are vertical angles. Therefore, if $b + c = e + f$ were true, this would mean that the measure of angle c would have the equal the measure of angle f. However, angles c and f are made with two different transversals, so there is no way to prove that this is true.

84. b. $x \le -\frac{17}{2}$ or $x \ge \frac{23}{2}$. With absolute value inequalities, you need to solve two equations. First, solve the original equation as is, but without the absolute value signs. Then, for the second equation, make one side negative and switch the direction of the inequality before solving.

$2x - 3 \ge 20$ results in $x \ge \frac{23}{2}$

$2x - 3 \le -20$ results in $x \le -\frac{17}{2}$

85. j. $(32x^3 + 8x^2)\pi$. The volume of the cylinder is the (area of the circular base) × (height).

$V = (\pi r^2)(height) = (\pi(2x)^2)(8x + 2) = (\pi 4x^2)(8x + 2) = \pi(32x^3 + 8x^2)$.

86. d. Sets A and B are disjoint. Sets A and B have no numbers in common, so they are disjoint.

87. g. similar. Isosceles right triangles are always 45-45-90 right triangles, so by angle-angle-angle Daryl will know that they are similar.

88. c. 48°. $\angle y$ is a corresponding angle to the 48° angle with Main Street and Jackson Street.

89. j. $\frac{25}{\pi}$. Circumference = $2\pi r = 10$, so $r = \frac{5}{\pi}$. Area = $\pi r^2 = \pi(\frac{5}{\pi})^2 = \frac{25}{\pi}$.

90. b. $q < |r| < |p|$. Assuming that q and r are half of $-\frac{1}{3}$ and $\frac{1}{3}$ respectively, we can estimate that $q = (-\frac{1}{3})(\frac{1}{2}) = -\frac{1}{6}$ and $r = \frac{1}{6}$. $P = -\frac{1}{2}$. It stands that $q < |r| < |p|$ is true since $-\frac{1}{6} < \frac{1}{6} < \frac{1}{2}$.

91. k. 5. Simplify $y - 4 = 5(x + 7)$ by getting y alone: $y = 4 + 5x + 35 = 5x + 39$. The slope is 5.

92. a. sector EOD. If the radius of the circle is 12 units, the area of the full circle is $\pi r^2 = 144\pi$. An area of 24π square units represents 16.67% of the circle since $\frac{24\pi}{144\pi} = 16.67\%$. Sector EOD represents 16.67% of the circle, since $\frac{60°}{360°} = 16.67\%$.

93. h. 25π **square units.** If the radius of the circle is 15 units, the area of the full circle is $\pi r^2 = 225\pi$. Sector DOB represents 11.11% of the circle since $\frac{40°}{360°} = 11.11\%$. 11.11% of $225\pi = (0.1111)(225\pi) = 25\pi$.

94. e. p **must be divisible by 80.** p must be divisible by all factors of both 20 and 4. 80 is a *multiple* of 20 and 4, but not a *factor* of 20 and 4.

95. j. 120. An inscribed angle is always half the measure of the arc it defines. The 60-degree angle defines an arc of 120 degrees and the x-degree angle defines the remaining arc in the circle which is 360 − 120 = 240 degrees. That 240-degree arc defines an angle that is 120 degrees and that angle is x.

CHAPTER

8 ▶ PRACTICE TEST 6

The *SHSAT Power Practice* tests will help you prepare for the high-stakes exams given to students applying for New York City's specialized high schools. Each practice test consists of sample questions like those you will find on the official SHSAT.

The 45-question verbal section and 50-question math section were developed by education experts. These tests will show you how much you know and what kinds of problems you still need to study. Mastering these practice tests will allow you to reach your highest potential on the real SHSAT.

PART I VERBAL

Scrambled Paragraphs

Paragraph 1
(q) (r) (s) (t) (u)
(q) (r) (s) (t) (u)
(q) (r) (s) (t) (u)
(q) (r) (s) (t) (u)
(q) (r) (s) (t) (u)

Paragraph 2
(q) (r) (s) (t) (u)
(q) (r) (s) (t) (u)
(q) (r) (s) (t) (u)
(q) (r) (s) (t) (u)
(q) (r) (s) (t) (u)

Paragraph 3
(q) (r) (s) (t) (u)
(q) (r) (s) (t) (u)
(q) (r) (s) (t) (u)
(q) (r) (s) (t) (u)
(q) (r) (s) (t) (u)

Paragraph 4
(q) (r) (s) (t) (u)
(q) (r) (s) (t) (u)
(q) (r) (s) (t) (u)
(q) (r) (s) (t) (u)
(q) (r) (s) (t) (u)

Paragraph 5
(q) (r) (s) (t) (u)
(q) (r) (s) (t) (u)
(q) (r) (s) (t) (u)
(q) (r) (s) (t) (u)
(q) (r) (s) (t) (u)

Logical Reasoning

6. (a) (b) (c) (d) (e)
7. (f) (g) (h) (j) (k)
8. (a) (b) (c) (d) (e)
9. (f) (g) (h) (j) (k)
10. (a) (b) (c) (d) (e)
11. (f) (g) (h) (j) (k)
12. (a) (b) (c) (d) (e)
13. (f) (g) (h) (j) (k)
14. (a) (b) (c) (d) (e)
15. (f) (g) (h) (j) (k)

Reading

16. (a) (b) (c) (d) (e)
17. (f) (g) (h) (j) (k)
18. (a) (b) (c) (d) (e)
19. (f) (g) (h) (j) (k)
20. (a) (b) (c) (d) (e)
21. (f) (g) (h) (j) (k)
22. (a) (b) (c) (d) (e)
23. (f) (g) (h) (j) (k)
24. (a) (b) (c) (d) (e)
25. (f) (g) (h) (j) (k)
26. (a) (b) (c) (d) (e)
27. (f) (g) (h) (j) (k)
28. (a) (b) (c) (d) (e)
29. (f) (g) (h) (j) (k)
30. (a) (b) (c) (d) (e)

31. (f) (g) (h) (j) (k)
32. (a) (b) (c) (d) (e)
33. (f) (g) (h) (j) (k)
34. (a) (b) (c) (d) (e)
35. (f) (g) (h) (j) (k)
36. (a) (b) (c) (d) (e)
37. (f) (g) (h) (j) (k)
38. (a) (b) (c) (d) (e)
39. (f) (g) (h) (j) (k)
40. (a) (b) (c) (d) (e)
41. (f) (g) (h) (j) (k)
42. (a) (b) (c) (d) (e)
43. (f) (g) (h) (j) (k)
44. (a) (b) (c) (d) (e)
45. (f) (g) (h) (j) (k)

PART II MATHEMATICS

46. (a) (b) (c) (d) (e)
47. (f) (g) (h) (j) (k)
48. (a) (b) (c) (d) (e)
49. (f) (g) (h) (j) (k)
50. (a) (b) (c) (d) (e)
51. (f) (g) (h) (j) (k)
52. (a) (b) (c) (d) (e)
53. (f) (g) (h) (j) (k)
54. (a) (b) (c) (d) (e)
55. (f) (g) (h) (j) (k)
56. (a) (b) (c) (d) (e)
57. (f) (g) (h) (j) (k)
58. (a) (b) (c) (d) (e)
59. (f) (g) (h) (j) (k)
60. (a) (b) (c) (d) (e)
61. (f) (g) (h) (j) (k)
62. (a) (b) (c) (d) (e)

63. (f) (g) (h) (j) (k)
64. (a) (b) (c) (d) (e)
65. (f) (g) (h) (j) (k)
66. (a) (b) (c) (d) (e)
67. (f) (g) (h) (j) (k)
68. (a) (b) (c) (d) (e)
69. (f) (g) (h) (j) (k)
70. (a) (b) (c) (d) (e)
71. (f) (g) (h) (j) (k)
72. (a) (b) (c) (d) (e)
73. (f) (g) (h) (j) (k)
74. (a) (b) (c) (d) (e)
75. (f) (g) (h) (j) (k)
76. (a) (b) (c) (d) (e)
77. (f) (g) (h) (j) (k)
78. (a) (b) (c) (d) (e)
79. (f) (g) (h) (j) (k)

80. (a) (b) (c) (d) (e)
81. (f) (g) (h) (j) (k)
82. (a) (b) (c) (d) (e)
83. (f) (g) (h) (j) (k)
84. (a) (b) (c) (d) (e)
85. (f) (g) (h) (j) (k)
86. (a) (b) (c) (d) (e)
87. (f) (g) (h) (j) (k)
88. (a) (b) (c) (d) (e)
89. (f) (g) (h) (j) (k)
90. (a) (b) (c) (d) (e)
91. (f) (g) (h) (j) (k)
92. (a) (b) (c) (d) (e)
93. (f) (g) (h) (j) (k)
94. (a) (b) (c) (d) (e)
95. (f) (g) (h) (j) (k)

Part 1—Verbal

The Verbal Test includes 45 questions in these three sections:

- Scrambled Paragraphs, 5 paragraphs (each counts double)
- Logical Reasoning, 10 questions, numbered 6–15
- Reading, 30 questions, numbered 16–45

Scrambled Paragraphs

This section tests your ability to organize a paragraph well. There are five paragraphs, presented in scrambled order. Your job is to put them in the best order to make a clear, coherent paragraph. Each correct answer counts double; these five paragraphs are worth 10 points out of the 50-point verbal test.

The first sentence in each paragraph is given. The remaining five sentences are listed in random order. Read each group of sentences carefully, and then decide on the best arrangement for them. Use the blanks at the left of each sentence to number these sentences from 1 to 5, showing the order they should be in.

Paragraph 1

O'Connell Street in Dublin City is named for Daniel O'Connell, an Irish patriot whose monument stands at one end of the street.

_____ **Q.** But witty Dubliners will not relinquish bragging rights easily.

_____ **T.** They will trump the French visitor with a fine distinction: The Champs Elysees is the widest *boulevard*, but O'Connell is the widest *street*.

_____ **S.** This claim usually meets with protests, especially from French tourists who boast that the Champs Elysees of Paris is Europe's widest street.

_____ **U.** It is Dublin's main thoroughfare.

_____ **R.** Although this thoroughfare is not a particularly long street, Dubliners will proudly tell visitors that it is the widest street in all of Europe.

Paragraph 2

Although table forks are perfectly ordinary objects for most people today, this eating utensil has a very long and interesting history.

_____ **Q.** During the tenth to thirteenth centuries, the practice of using a fork as a personal eating utensil began to spread across Europe.

_____ **R.** Historians believe that some of the first table forks were used in ancient Greece.

_____ **S.** Middle Easterners adopted the use of the fork around 7 AD.

_____ **T.** After Greece, personal table forks made their way to France and Italy, but they didn't become popular in these countries until sometime in the 1500s or 1600s.

_____ **U.** It quickly became popular for wealthy Greeks to use the fork in this manner.

Paragraph 3

Table forks were brought to England in the early 1600s.

_____ **Q.** As forks became more popular, English craftspeople began making them out of precious metals and with intricate patterns that were designed to impress visitors.

_____ **R.** The noblemen rejected forks as unmanly because, at that time, it was customary to eat with one's hands.

_____ **S.** By the mid–1600s, elegant forks were considered a symbol of refinement and status, and were popular among English nobility.

_____ **T.** At first, English noblemen made fun of the new eating utensils.

_____ **U.** In spite of that first reaction among noblemen, the fork slowly caught on in England among the wealthy.

Paragraph 4

One of the most hazardous conditions a firefighter will ever encounter is a backdraft (also known as a smoke explosion).

_____ **Q.** Those other warning signs include little or no visible flame, excessive heat, smoke leaving the building in puffs, muffled sounds, and smoke-stained windows.

_____ **R.** When there is a lack of oxygen during a fire, the smoke becomes filled with carbon dioxide or carbon monoxide and turns dense gray or black.

_____ **S.** A backdraft can occur in the hot, smoldering phase of a fire when burning is incomplete and there is not enough oxygen to sustain the fire.

_____ **T.** Then, if more oxygen reaches the fire, the intense heat combined with unburned carbon and other flammable products can cause powerful combustion.

_____ **U.** That's why firefighters must pay careful attention to dark smoke and other signs of a potential backdraft.

Paragraph 5

On February 3, 1956, Autherine Lucy became the first African-American student to attend the University of Alabama.

_____ **Q.** When she did, white students rioted to protest her admission, and the federal government had to assume

command of the Alabama National Guard in order to protect her.

_____ **R.** Later she would appreciate his seeming indifference, as he was one of only a few professors to speak out in favor of her right to attend the university.

_____ **S.** In spite of the unfriendly dean and the rioting students, Autherine bravely took a seat in the front row on her first day in class.

_____ **T.** She remembers being surprised that the professor appeared not to notice she was in class that first day.

_____ **U.** Even though the dean of women refused to allow her to live in a university dormitory, Autherine still wanted to attend classes at the university.

Logical Reasoning

The questions in this section test your ability to reason well, that is, to figure out what the facts you know can or can't possibly mean. Read the statements carefully, then choose the best answer based *only* on the information given. Note carefully the words used in each question. For example, one thing can be larg*er* than another without being the larg*est* in the group. In answering some of these questions, it may be useful to draw a rough diagram or make a list that gives real world values to the information.

6. Mary is taking a bus ride from New York to Boston. The bus ride from New York to Boston takes approximately 4.5 hours.

   Which of the following can we assume is true?

   a. Mary does not enjoy her bus ride.
   b. It takes approximately 4.5 hours by bus to get from New York to Boston.
   c. Mary will be in Boston in 4.5 hours.
   d. Mary has never taken a bus ride to Boston before.
   e. The return trip from Boston to New York will take approximately 5 hours.

7. Occasionally, Jonathan and Alberto go to used vinyl stores together to search for rare, out-of-print albums. Today was a great day for both of them—Jonathan found three albums and Alberto found two.

   f. Jonathan and Alberto are very knowledge-able about rare, out-of-print albums.
   g. Alberto found the rarest album he had yet discovered.
   h. Jonathan usually finds better deals then Alberto.
   j. One of the albums that Jonathan found was much too expensive.
   k. Alberto found fewer albums than Jonathan.

8. Read the following statements.
   - Lake Mead is colder than Walden Pond.
   - Brown River is warmer than Red Swamp.
   - Red Swamp is colder than Lake Mead.

   If the first two statements are true, the third is

   a. true.
   b. false.
   c. partly true.
   d. repetitive.
   e. not possible to determine from the information given.

9. The only convenience store near Kevin's house is 20 minutes away and closes at midnight. At 11:50 P.M., Kevin remembered that he forgot to pick up a bottle of juice. He went to the store as soon as he remembered, but shortly returned empty-handed.

   Which of the following is most likely true?

   f. Kevin does not like going to the convenience store.
   g. The convenience store was probably already closed.
   h. Kevin did not go to the store.
   j. Kevin is usually very lazy.
   k. It is impossible to determine from the information given.

Read the following passage, then answer Questions 10 through 12.

10. Five cities all got more rain than usual this year. The five cities are: Last Stand, Mile City, New Town, Olliopolis, and Polberg. The cities are located in five different areas of the country: in the mountains, the forest, the coast, the desert, and in a valley. The rainfall amounts were: 12 inches, 27 inches, 32 inches, 44 inches, and 65 inches.

Read the following facts carefully:
- The city in the desert got the least rain; the city in the forest got the most rain.
- New Town is in the mountains.
- Last Stand got more rain than Olliopolis.
- Mile City got more rain than Polberg, but less rain than New Town.
- Olliopolis got 44 inches of rain.
- The city in the mountains got 32 inches of rain; the city on the coast got 27 inches of rain.

Which city is in the desert?

- **a.** Last Stand
- **b.** Mile City
- **c.** New Town
- **d.** Olliopolis
- **e.** Polberg

11. How much rain did Mile City get?
- **f.** 12 inches
- **g.** 27 inches
- **h.** 32 inches
- **j.** 44 inches
- **k.** 65 inches

12. Where is Olliopolis located?
- **a.** in a valley
- **b.** the coast
- **c.** the mountains
- **d.** the desert
- **e.** the forest

13. Read the following statements:
- All the houses on Reynolds Road have roofs made of shingles.
- No shingles are purple.
- None of the houses on Reynolds Road have purple roofs.

If the first two statements are true, the third is

- **f.** true.
- **g.** false.
- **h.** probably false.
- **j.** partly true.
- **k.** not possible to determine from the information given.

14. Before going on an out-of-town business trip, Natasha took her cats over to her friend Amy's house. Shortly after Natasha's departure, Amy got ill and had to be taken to a hospital, where she stayed for the next two days.

Which of the following must be true?

- **a.** Natasha's cats went hungry for two days.
- **b.** When Natasha learned that Amy was in a hospital, she canceled her trip.
- **c.** Natasha picked up her cats after two days.
- **d.** Amy felt bad about leaving Natasha's cats.
- **e.** It cannot be determined from the information given.

15. May keeps her spare change in a bowl on her desk. A dollar is missing from her bowl. May's brother Andy is always borrowing change from her.

Which of the following can we conclude is true?

f. Andy does not have an income.

g. Andy took a dollar from May's bowl.

h. May always has a lot of spare change.

j. May keeps her spare change in a bowl on her desk.

k. It cannot be determined from the information given.

Reading Comprehension

This section tests your reading comprehension— your ability to understand what you read. Read each passage carefully and answer the questions that follow it. If necessary, you can reread the passage to be certain of your answers. Remember that your answers must be based only on information that is actually in the passage.

Read the following passage, then answer Questions 16 through 20.

During those barren winter months, with windows overlooking long-dead gardens, leafless trees, and lawns that seem to have an ashy look about them, nothing soothes the jangled nerves more than the vibrant green of plants surrounding the living spaces of one's home. People browse through garden stores just to get a whiff of chlorophyll and to choose a plant or two to bring spring back into their winter-gray lives.

Now there is even more of a need for "the green," in light of recent articles warning us of the hazards of chemicals that we, ourselves, introduce into our homes. Each time we bring clothes home from the cleaners, we release those chemicals into the closed-in air of our dwellings. Every cleanser releases its own assortment of fumes. Some of the chemicals are formaldehyde, chlorine, benzene, and styrene. Read the labels on many home products; the ingredients aren't even listed!

During the winter, when those same windows are shut tight, we breathe in these chemicals—causing symptoms much like allergies. In fact, most people probably dismiss the effects of these chemicals simply as a flare-up of an allergy. The truth is that we are experiencing a syndrome that is called Multiple Chemical Sensitivity. Now, what has this got to do with green plants? Everything healthy! Research has been conducted with two types of plants that have actually removed a great deal of these harmful chemicals from the air.

The two plants that seem to be the best bet for ridding one's home of such chemicals are ferns and palms. These plants release moisture as part of photosynthesis and, as they do, pull chemicals from the air into their leaves. Even NASA has conducted some greenhouse experiments for long-term space exploration. Within hours, their plants (palms) had removed almost all traces of formaldehyde in the room.

Both ferns and palms are ancient species of plants, dating back more than a hundred million years. Another trait they share is that they both live long lives, 100 years or more. This we expect from trees, but ferns and palms are plants—plants that can grow to 65 feet in the proper setting! Even their individual leaves live for one to two years (ferns) and one to nine years (palms). Perhaps it is their primal qualities that have contributed to their ability to purify their environment.

16. Which of the following best tells what this passage is about?
- **a.** Our homes are full of contaminants.
- **b.** Our allergies are caused by chemicals found in the home.
- **c.** All plants release moisture in the home.
- **d.** Certain plants can purify the home of many harmful chemicals.
- **e.** Plants help cheer up a room in the winter.

17. According to the passage, when a few harmful chemicals combine, they can
- **f.** cause us to experience allergies.
- **g.** cause a monumental task for homeowners.
- **h.** cause houseplants to die.
- **j.** contribute to photosynthesis in plants.
- **k.** contribute to a syndrome called Multiple Chemical Sensitivity.

18. The passage indicates that research
- **a.** has only been conducted using specific plants.
- **b.** has only been conducted by NASA.
- **c.** has not identified the sources of these chemical impurities.
- **d.** has only benefited long-term space exploration.
- **e.** has only been conducted in a few house-holds.

19. The passage infers a relationship between the antiquity of ferns and palms and their ability to
- **f.** live long.
- **g.** purify the air.
- **h.** reproduce rapidly.
- **j.** react successfully in research experiments.
- **k.** grow leaves that live long.

Read the following passage, then answer Questions 20 through 24.

The little flat mail-pockets strapped under the rider's thighs would each hold about the bulk of a child's primer [school book]. They held many an important business chapter and newspaper letter, but these were written on paper as airy and thin as goldleaf, nearly, and thus bulk and weight were economized. The stagecoach traveled about a hundred to a hundred and twenty-five miles a day (twenty-four hours), the pony-rider about two hundred and fifty. There were about eighty pony-riders in the saddle all the time, night and day, stretching in a long, scattering procession from Missouri to California, 40 flying eastward, and 40 toward the west, and among them making 400 gallant horses earn a stirring livelihood and see a deal of scenery every single day in the year.

We had a consuming desire, from the beginning, to see a pony-rider, but somehow or other all that passed us and all that met us managed to streak by in the night, and so we heard only a whiz and a hail, and the swift phantom of the desert was gone before we could get our heads out of the windows. But now we were expecting one along every moment, and would see him in broad daylight. Presently the driver exclaims:

"HERE HE COMES!"

Every neck is stretched further, and every eye strained wider. Away across the endless dead level of the prairie a black speck appears against the sky, and it is plain that it moves. Well, I should think so! In a second or two it becomes a horse and rider, rising and falling, rising and falling, rising and falling—sweeping toward us nearer and nearer—growing more and more distinct, more and more sharply defined—nearer and still nearer, and the flutter

of the hoofs comes faintly to the ear—another instant a whoop and a hurrah from our upper deck, a wave of the rider's hand, but no reply, and a man and a horse burst past our excited faces, and go swinging away like a belated fragment of a storm!
—from *Roughing It* by Mark Twain

20. Which of the following best tells what this passage is about?

   a. A Pony Express rider relates how he delivered the mail in the old West.

   b. The narrator explains the Pony Express system and describes seeing a rider go by.

   c. A stagecoach driver describes seeing a Pony Express rider race past the coach.

   d. A writer tells about his unfulfilled desire to see one of the famous Pony Express riders.

   e. While traveling westward by train, the narrator sees a Pony Express rider go by.

21. The tone of the passage suggests that the author's attitude toward the Pony Express rider is

   f. indifference.

   g. fear.

   h. bewilderment.

   j. excitement.

   k. annoyance.

22. The narrator of the passage is most likely

   a. sitting alone on a porch.

   b. inside a stagecoach with other passengers.

   c. riding in a hot air balloon.

   d. picnicking atop a hill with friends.

   e. driving a covered wagon.

23. Which of the following is NOT supported by the passage?

   f. The mail was strapped in a pouch under the rider's thighs.

   g. The rider rode great distances to deliver the mail.

   h. Four hundred horses carried Pony Express riders every day.

   j. Usually eighty pony riders were in the saddle at any given time.

   k. In one day, a stagecoach could travel twice as far as a pony rider.

24. The author and his companions had never seen a Pony Express rider before because

   a. the Pony Express mail service had just gotten started.

   b. the pony riders avoided the main roads and cut across rough country.

   c. most pony riders never made it that far.

   d. the pony riders had only passed them at night.

   e. they were bored and didn't bother to look up as the pony riders raced by.

Read the following passage, then answer Questions 25 through 30.

Gordon Parks's first professional photograph, which he titled "American Gothic" after a well-known painting by Grant Wood, remains one of the most powerful images of minorities created in the years leading up to the Civil Rights Movement. The composition of Parks's photograph echoes that of Wood's painting, but while the painting depicts a rigidly stoic farmer and his wife holding a pitchfork against the backdrop of a rustic American farm, the photograph shows an African-American charwoman (a woman hired to clean) holding a broom and mop against the backdrop of a looming American flag. The woman's gaze is direct yet not accusatory; her glance is slightly averted, suggesting a natural tendency to avoid confrontation. The power of the image comes from the contrast of the subject's almost painfully deep expression with the glorious ideals symbolized by the American flag.

Parks captured this memorable image in 1942 on his first day working for the Farm Security Administration (FSA). This government agency was established by President Franklin D. Roosevelt to aid workers during the harsh years of the Depression. Photographers like Parks were hired to create images that could communicate the plight of these Americans. As an African American, Parks had many firsthand experiences of discrimination and prejudice. Yet he was still surprised by the attitudes he encountered when he began working for the FSA in Washington, DC. He was forced to enter restaurants from rear entrances and forbidden from even going into some theaters and other locations. Parks's new boss, Roy Stryker, suggested that he talk with some African-American residents in Washington and use his camera to record their experiences.

Parks followed his mentor's advice, which led him to Ella Watson, the subject of "American Gothic." Parks met her while she was cleaning floors in the FSA building. She told him of her life, which included many instances of bigotry and extreme hardship. When she agreed to let the young photographer take her picture, Watson stepped into one of the most recognized and influential photographs of our time. But when Parks brought the photograph to Stryker, he was both impressed and cautious.

"You're getting the idea," Stryker commented, "but you're going to get us all fired."

Parks accepted Stryker's appraisal, recognizing that the photograph might be too sensational for large-scale publication. So it was both surprise and pride that Parks felt when he saw "American Gothic" on the front page of the *Washington Post*. Ella Watson's face spoke volumes about her past, and Parks's piercing composition added an unavoidable editorial commentary.
—Stanley Isaacs

25. Which of the following best tells what this passage is about?
    a. why Gordon Parks decided to become a professional photographer
    b. why Ella Watson took a job sweeping floors
    c. how President Franklin D. Roosevelt tried to aid workers during of harsh years of the depression
    d. how Gordon Parks faced discrimination
    e. why a photograph of Ella Watson taken by Gordon Parks became important

26. Which of the following is most likely true of Gordon Parks's career?
    f. Gordon Parks became famous for photographing celebrities.
    g. Gordon Parks made a career out of making photographs of existing artwork.
    h. Gordon Parks became famous for his photographs of ordinary working people.
    j. Gordon Parks had a long career in fashion photography.
    k. Gordon Parks is famous for his photographs of landscapes in the American West.

27. Which of the following statements about Gordon Parks's photograph "American Gothic" is NOT true according to the passage?
   a. It was his first professional photograph.
   b. The picture was of a group of African-American residents in Washington.
   c. In the background was an American flag.
   d. The image was considered sensational at that time.
   e. It ran on the front page of the *Washington Post*.

28. Why did Gordon Parks choose the name "American Gothic" for his photograph?
   f. The photograph was dark and mysterious.
   g. The photograph exactly duplicated the images of Grant Wood's painting with the same name.
   h. The photograph had a composition similar to Grant Wood's painting.
   j. He admired Grant Wood's life and work.
   k. Ella Watson bore an uncanny resemblance to the farmer's wife in Grant Wood's painting.

29. Why did the Farm Security Administration hire photographers like Gordon Parks?
   a. to create advertisements that would recruit new farmers
   b. to produce photographs showing the difficult lives of American workers
   c. to take photographs that could be sold as souvenirs or mementos
   d. to make photojournalism more influential than print journalism
   e. to provide direct aid to struggling farmers and other workers

30. Talking about his career choices, Parks has said, "I picked up a camera because it was my choice of weapons against what I hated most about the universe: racism, intolerance, poverty." Using what you know from the excerpt and this quotation, why does Parks say that a camera is a "weapon"?

   Parks believes that a camera

   f. creates pictures that often get people fired.
   g. is a tool an artist uses to capture beauty and light.
   h. can capture and destroy the soul of people who are photographed.
   j. is dangerous because it can be used to create propaganda.
   k. can create images that lead to social and political change.

Read the following passage, then answer Questions 31 through 34.

Dennis Hope, self-proclaimed Head Cheese of the Lunar Embassy, will promise you the moon. Or at least a piece of it. Since 1980, Hope has raked in over $9 million selling acres of lunar real estate for $19.99 a pop. So far, 4.25 million people have purchased a piece of the moon, including celebrities like Barbara Walters, George Lucas, Ronald Reagan, and even the first President Bush.

Hope says he exploited a loophole in the 1967 United Nations Outer Space Treaty, which prohibits nations from owning the moon. Because the law says nothing about individual holders, he says, his claim—which he sent to the United Nations—has some clout. "It was unowned land," he says. "For private property claims, 197 countries at one time or another had a basis by which private citizens could make claims on land and not make payment. There are no standardized rules."

Hope is right that the rules are somewhat murky—both Japan and the United States have plans for moon colonies—and lunar property ownership might be a powder keg waiting to spark. But Ram Jakhu, law professor at the Institute of Air and Space Law at McGill University in Toronto, says that Hope's claims aren't likely to hold much weight. Nor, for that matter, would any nation's. "I don't see a loophole," Jakhu says. "The moon is a common property of the international community, so individuals and states cannot own it. That's very clear in the U.N. treaty. Individuals' rights cannot prevail over the rights and obligations of a state."

Jakhu, a director of the International Institute for Space Law, believes that entrepreneurs like Hope have misread the treaty and that the 1967 legislation came about to block property claims in outer space. Historically, "the ownership of private property has been a major cause of war," he says. "No one owns the moon. No one can own any property in outer space."

Hope refuses to be discouraged. And he's focusing on expansion. "I own about 95 different planetary bodies," he says. "The total amount of property I currently own is about 7 trillion acres. The value of that property is about $100 trillion. And that doesn't even include mineral rights."

—"Location, Location, Location, Who Owns the Moon?" by Stephen Ornes. Reprinted with permission from the July 2007 issue of *Discover* magazine. Copyright © *Discover* Magazine. All rights reserved.

31. Which of the following best tells what this passage is about?
   a. the large amount of money that Dennis Hope has made by selling land on the moon
   b. the ownership of private property has been a major cause of war
   c. individual rights are more important than the rights of the state
   d. the rights of the state are more important than individual rights
   e. the argument over whether Dennis Hope has the right to sell land on the moon

32. Which of the following statements best summarizes Dennis Hope's argument for his right to own and sell property on the moon?
   f. Celebrities have a right to own moon property.
   g. The treaties do not say that individuals cannot own land in outer space.
   h. The moon is common property of the international community.
   j. A lot of money can be made by selling property in outer space.
   k. Individuals' rights cannot prevail over the rights of a state.

33. According to the author of this passage, what might possibly give Hope the right to sell property on the moon?
   a. Hope refuses to be discouraged.
   b. Hope has raked in over $9 million selling acres of lunar real estate.
   c. Hope is right; the rules are somewhat murky.
   d. Space is running out on the moon so people should buy now.
   e. Hope has a great idea and should be able to go ahead and sell land.

34. Why does the author include the detail about the number of people and the list of celebrities who have bought pieces of land on the moon?

  f. to show that buying land on the moon is silly

  g. to show that Dennis Hope has made a lot of money

  h. to prove that Dennis Hope has the right to sell property on the moon

  j. to show that owning land on the moon has become popular

  k. to show that space treaties have many loopholes

Read the following passage, then answer Questions 35 through 39.

For centuries, time was measured by the position of the sun with the use of sundials. Noon was recognized when the sun was the highest in the sky, and individual cities would set their clocks by this apparent solar time. Daylight saving time (DST), sometimes called summer time, was instituted to make better use of daylight. Thus, clocks are set forward one hour in the spring to move an hour of daylight from the morning to the evening and then set back one hour in the fall to return to normal daylight.

Benjamin Franklin first conceived the idea of daylight saving time during his tenure as an American delegate in Paris in 1784. It is said that Franklin awoke early one morning and was surprised to see the sunlight at such an hour. Always the economist, Franklin believed the practice of moving the time could save on the use of candlelight, as candles were expensive.

In England, builder William Willett (1857–1915) became a strong supporter for daylight saving time upon noticing blinds of many houses were closed on an early sunny morning. Willett believed everyone, including himself, would appreciate longer hours of light in the evenings. In 1909, Sir Robert Pearce introduced a bill in the House of Commons to make it obligatory to adjust the clocks. A bill was drafted and introduced into Parliament several times but met with great opposition, mostly from farmers. Eventually, in 1925, it was decided that summer time should begin on the day following the third Saturday in April and close after the first Saturday in October.

The U.S. Congress passed the Standard Time Act of 1918 to establish standard time and preserve and set daylight saving time across the continent. This act also devised five time zones throughout the United States: Eastern, Central, Mountain, Pacific, and Alaska.

President Roosevelt established year-round daylight saving time (also called war time) from 1942 to 1945. However, after this period, each state adopted its own DST, which proved to be disconcerting to television and radio broadcasting and transportation. In 1966, President Lyndon Johnson created the Department of Transportation and signed the Uniform Time Act. As a result, the Department of Transportation was given the responsibility for the time laws.

During the oil embargo and energy crisis of the 1970s, President Richard Nixon extended DST through the Daylight Saving Time Energy Act of 1973 to conserve energy further. This law was modified in 1986, and daylight saving time was reset to begin on the first Sunday in April (to spring ahead) and end on the last Sunday in October (to fall back).

35. Which of the following titles best tells what this passage is about?

    a. The History and Rationale of Daylight Saving Time

    b. Lyndon Johnson and the Uniform Time Act

    c. The U.S. Department of Transportation and Daylight Saving Time

    d. Daylight Saving Time in the United States

    e. Benjamin Franklin and Daylight Saving Time

36. Who first thought up the idea of DST?

    f. President Richard Nixon

    g. Benjamin Franklin

    h. Sir Robert Pearce

    j. President Lyndon Johnson

    k. William Willett

37. Who opposed the bill that was introduced in the House of Commons in the early 1900s?

    a. Sir Robert Pearce

    b. farmers

    c. television and radio broadcasting companies

    d. the U.S. Congress

    e. families with school-age children

38. Which of the following statements is true of the U.S. Department of Transportation?

    f. It was created by President Richard Nixon.

    g. It set standards for DST throughout the world.

    h. It constructed the Uniform Time Act.

    j. It oversees all time laws in the United States.

    k. It was first established between 1942 and 1945.

39. The Daylight Saving Time Energy Act of 1973 was responsible for

    a. preserving and setting daylight saving time across the continent.

    b. instituting five time zones in the United States.

    c. extending daylight saving time in the interest of energy conservation.

    d. conserving energy by giving the Department of Transportation authority over time laws.

    e. honoring Benjamin Franklin's contributions to DST.

Read the following passage, then answer Questions 40 through 45.

Susan Brownell Anthony, a pioneer leader for women's rights, was born and raised in Massachusetts. Her parents were stern people who taught their children to have self-discipline, strong convictions, and belief in their own worth. One brother was active in the antislavery movement, and a sister became a teacher and activist for women's rights. The second-oldest child in the family, Susan, would become one of the leading women reformers of the nineteenth century.

In her first job as a schoolteacher, Susan soon realized that the male teachers were earning about four times what female teachers earned for doing the same job. She argued for equal wages for women, but finally left teaching and moved to Rochester, New York. In Rochester, Susan B. Anthony began her first public crusade on behalf of temperance (abstinence from alcoholic drinks). The temperance movement dealt with the abuses of women and children who suffered from alcoholic husbands. Also, she worked tirelessly against slavery and for women's rights. Anthony helped write the history of woman suffrage (the right to vote).

At the time that Anthony lived, women did not have the right to vote. When Anthony did vote in the 1872 election, a U.S. Marshall arrested her. She hoped to prove that women had the legal right to vote under the provisions of the fourteenth and fifteenth amendments to the Constitution. At her trial, a hostile federal judge found her guilty and fined her $100, which she refused to pay.

Anthony did not work alone. She collaborated with reformers working for women's rights, including Elizabeth Cady Stanton and Amelia Bloomer. Susan B. Anthony also worked for the American Anti-Slavery Society with Frederick Douglass, a fugitive slave and black abolitionist.

Although Anthony did not live to see the fruits of her efforts, she was among those who first proposed, in 1848, an amendment granting women the right to vote. The nineteenth amendment to the U.S. Constitution, ratified in 1920, finally granted women that right in all state and national elections. On July 2, 1979, the U.S. Mint honored her work by issuing the Susan B. Anthony $1 coin.

40. Which of the following best tells what this passage is about?
- **a.** Reformers do not always see the results of their efforts.
- **b.** Susan B. Anthony never gave up her fight for all people's freedoms.
- **c.** Slavery was one of Susan B. Anthony's causes.
- **d.** Anthony did not condone the use of alcohol.
- **e.** Anthony collaborated with Elizabeth Cady Stanton and Amelia Bloomer.

41. Anthony advocated all the following EXCEPT
- **f.** the abolition of slavery.
- **g.** prohibition of alcohol because of the abuse it causes.
- **h.** women are citizens and should have the right to vote.
- **j.** employers should provide child care for female employees.
- **k.** collaborating with other women who worked for the same causes.

42. According to the passage, Susan B. Anthony was an effective reformer because she
- **a.** had the support of a judge.
- **b.** enlisted the help of others to promote a cause.
- **c.** was well-informed about the causes.
- **d.** ignored what others thought.
- **e.** was bilingual.

43. Which of the following best explains what the author means by saying that Susan B. Anthony "began her first public crusade" in Rochester?
- **f.** She began a war against the infidels, like those in the Middle Ages.
- **g.** She began a quest to fight evil around the world.
- **h.** She began a battle against any kind of authority.
- **j.** She began a campaign to work tirelessly for her beliefs.
- **k.** She began a struggle against narrow-minded judges.

44. Which of the following achievements does the passage about Susan B. Anthony emphasize most?

    a. She collaborated with abolitionists to rid the country of slavery.

    b. She was an activist and raised a family at the same time.

    c. Her tireless efforts led to the establishment of the nineteenth amendment.

    d. She was a leader in the temperance movement.

    e. She worked for justice in the world.

45. In which of the following ways did the U.S. Mint honor her life's work?

    f. The Susan B. Anthony stamp was issued.

    g. The Susan B. Anthony dollar was created.

    h. Susan B. Anthony board games were created and became popular.

    j. The Susan B. Anthony Memorial Park was built in Rochester.

    k. Susan B. Anthony dolls were created and became popular.

Part 2—Math

The Math Test includes 50 questions covering content in the following areas:

- basic math
- percentages, fractions, decimals, averages
- pre-algebra
- algebra
- substitution
- factoring
- geometry
- probability
- logic
- word problems

Solve each problem and select the best answer from the choices given. It is important to keep in mind that:

- Formulas and definitions of mathematical terms and symbols are not provided.
- Diagrams other than graphs are not necessarily drawn to scale. Do not assume any relationship in a diagram unless it is specifically stated or can be figured out from the information given.
- A diagram is in one plane unless the problem specifically states that it is not.
- Graphs are drawn to scale. Unless stated otherwise, you can assume relationships according to appearance. For example, (on a graph) lines that appear to be parallel can be assumed to be parallel; likewise for concurrent lines, straight lines, collinear points, right angles, and so on.
- You need to reduce all fractions to lowest terms.

46. Joel wants to purchase a CD for $16.25. The clerk tells him he must pay a sales tax equal to 8% of his purchase. How much sales tax must Joel pay?

    a. $1.30

    b. $1.63

    c. $2.03

    d. $2.18

    e. $13.00

47. In isosceles $\triangle ABC$, if vertex $\angle A$ is twice the measure of base $\angle B$, then what must $\angle C$ measure?

    f. 30°

    g. 33°

    h. 55°

    j. 45°

    k. 90°

48. This graph shows the number of inches of rain for 5 towns in Suffolk County during Spring 2009.

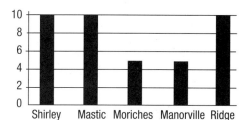

What was the mean number of inches for the season shown?

a. 5

b. 7.5

c. 8

d. 9

e. 10

49. In a seminar, the ratio of men to women is 5:9. If there are 56 people in the seminar, how many are women?

f. 4

g. 7

h. 20

j. 26

k. 36

50. Which of the following correctly shows the inequality $-1.4 \leq x \leq 2$?

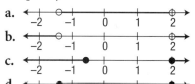

e. None of the above closely model $-1.4 \leq x \leq 2$.

51. A band recently sold 5,000 tickets to a concert. If there are two types of tickets, regular and children's, and there were 600 more children's tickets sold than regular tickets, how many children's tickets were sold?

f. 2,200

g. 2,500

h. 3,200

j. 2,800

k. 4,400

52. Pentagons *ABCDE* and *FGHIJ* are similar. The ratio of each side of pentagon *ABCDE* to its corresponding side of pentagon *FGHIJ* is 4:1. If $\overline{AB}$ and $\overline{FG}$ are corresponding sides, and the length of is $4x + 4$, what is the length of $\overline{FG}$?

a. $x + 1$

b. $2x + 2$

c. $4x + 4$

d. $16x + 16$

e. $20x + 20$

53. What is the area of a square that has a diagonal length of 14 m? (Hint: The Pythagorean theorem will be helpful.)

f. 56 m^2

g. 84 m^2

h. 196 m^2

j. 98 m^2

k. 294 m^2

54. If the coordinates of point *A* are $(-2,4)$ and the coordinates of point *B* are $(1,-3)$, what is the length of the line segment *AB* in radical form?

a. $\sqrt{58}$

b. $\sqrt{52}$

c. $5\sqrt{2}$

d. $\sqrt{10}$

e. $\sqrt{38}$

55. Terri fills $\frac{2}{3}$ of a glass with water. If the glass is 15 cm tall and its radius is 2 cm, what volume of water is in Terri's glass? Use the following formula for calculating the volume of a cylinder: $V = \pi r^2 \, (height)$.

   f. 10π cm^3

   g. 20π cm^3

   h. 30π cm^3

   j. 40π cm^3

   k. 50π cm^3

56. Carla's dance squad organizes a car wash in the municipal parking lot. It costs them $250 to rent the lot, and they pay $35 for cleansers. If the squad charges $5 per car wash, how many cars must they wash to raise more money than their expenses?

   a. 50 cars

   b. 51 cars

   c. 55 cars

   d. 57 cars

   e. 58 cars

57. Which equation represents a line that contains the point (0,–2) and is parallel to $\frac{1}{2}x + \frac{1}{4}y = \frac{1}{8}$?

   f. $y = 2x + \frac{1}{2}$

   g. $y = \frac{1}{2}x + \frac{1}{2}$

   h. $y = -2x - 2$

   j. $y = -2x + \frac{1}{2}$

   k. $y = -\frac{1}{2}x + 2$

58. 1,200 new nursing students were asked to complete a survey in which they were asked which type of nursing they would like to pursue. The data was used to make the pie chart below.

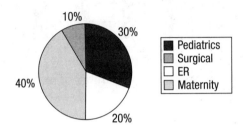

Nursing Survey

How many more nursing students wanted to pursue Maternity over ER?

   a. 480 students

   b. 120 students

   c. 600 students

   d. 300 students

   e. 240 students

59. If a drawer contains 7 navy socks, 4 white socks, and 9 black socks, what is the probability that the first sock randomly drawn out of the drawer will NOT be white?

   f. $\frac{1}{5}$

   g. $\frac{1}{4}$

   h. $\frac{7}{20}$

   j. $\frac{4}{5}$

   k. 16

60. A circle and a square have the same areas. The circle has a diameter of 8. What is the length of one of the square's sides?

   a. $4\sqrt{\pi}$

   b. $4\sqrt{2\pi}$

   c. 16π

   d. 4π

   e. $4\sqrt{6\pi}$

61. If z is even and v is odd, which of the following must be true?

   f. $2z > 0$

   g. v is prime.

   h. $3z$ is odd.

   j. z is a composite number.

   k. none of the above

62. Andrea draws a polygon with n number of sides. The sum of the interior angles of her polygon is 60 times its number of sides. How many sides does Andrea's polygon have?

   a. 3 sides

   b. 4 sides

   c. 5 sides

   d. 6 sides

   e. 10 sides

63. If angle 4 is congruent to angle 7, which of the following is NOT true?

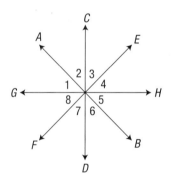

   f. angle 4 = angle 3

   g. angle 2 = angle 3

   h. angle 8 = angle 3

   j. angle 7 = angle 3

   k. angle 7 = angle 8

64. If the length of $\overline{AB}$ of square $ABCD$ is x units, and the length of $\overline{EF}$ is $\frac{2}{3}$ that of $\overline{AD}$, what is the size of the shaded area?

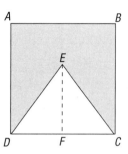

   a. $x^2 - \frac{1}{2}x$ square units

   b. $x^2 - \frac{1}{3}x$ square units

   c. $\frac{1}{3}x^2$ square units

   d. $x^2 - \frac{2}{3}x$ square units

   e. $\frac{2}{3}x^2$ square units

65. $(7x^3y^2)^2$

   f. $49x^3y^2$

   g. $49x^5y^4$

   h. $49x^6y^4$

   j. $7x^5y^4$

   k. $14x^6y^4$

66. Which of the following equations is correct?

   a. $\sqrt{36} + \sqrt{64} = \sqrt{100}$

   b. $\sqrt{25} + \sqrt{16} = \sqrt{41}$

   c. $\sqrt{9} + \sqrt{25} = \sqrt{64}$

   d. $\sqrt{16} + \sqrt{4} = \sqrt{20}$

   e. There is no correct equation.

67. In the diagram, if $\angle 1$ is 30° and $\angle 2$ is a right angle, what is the measure of $\angle 5$?

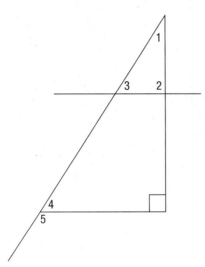

  f. 30°
  g. 60°
  h. 120°
  j. 140°
  k. 150°

68. If the lengths of the sides of a square are halved, the area of the new square is
  a. one-fourth the area of the old square.
  b. one-half the area of the old square.
  c. equal to the area of the old square.
  d. twice the area of the old square.
  e. four times the area of the old square.

69. If $p\&q$ is equivalent to $\frac{p}{q} + pq$, what is the value of $q\&p$ when $p = -2$ and $q = 4$?
  f. −10
  g. −8.5
  h. −7.5
  j. 4
  k. 16

70. If the prime factorization of f contains a 3 and the prime factorization of g contains a 2, then which of the following statements is NOT correct?
  a. $f \times g$ must be divisible by 4.
  b. $f \times g$ must be divisible by 3.
  c. $f \times g$ must be divisible by 2.
  d. $f \times g$ must be divisible by 6.
  e. $f \times g$ can be divisible by 5.

71. If $V = x^3 + y[r(3s + 2) + 4r]$ and $x = -2$, $y = 3$, $r = -5$, and $s = 4$, what is the value of V?
  f. −278
  g. 450
  h. 999
  j. 142
  k. −158

72. If the lacrosse team's total expenses this season were $5,400, how much more money was spent on transportation than on training?

Lacrosse Team Budget 2007

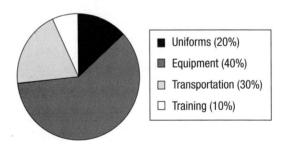

  a. $2,160
  b. $540
  c. $1,080
  d. $1,620
  e. $2,700

73. Given $f(x) = \frac{x^2}{5\sqrt{3}}$, what is the range of this function?

   f. all real numbers

   g. all real numbers greater than or equal to zero

   h. all real numbers except $x = 0$

   j. all real numbers greater than zero

   k. all perfect squares

74. Which of the following points is in the solution set of $4y + 6 > 3x + 15$?

   a. $(0,-4)$

   b. $(3,4)$

   c. $(4,3)$

   d. $(4,6)$

   e. $(5,6)$

75. Sammy fills $\frac{2}{7}$ of a bucket with sand. Jessie has an identical bucket and fills $\frac{3}{5}$ of it with sand. If Sammy pours all of the sand in his bucket into Jessie's bucket, what fraction of Jessie's bucket will be full?

   f. $\frac{5}{5}$

   g. $\frac{5}{12}$

   h. $\frac{21}{35}$

   j. $\frac{6}{35}$

   k. $\frac{31}{35}$

76. According to the diagram below, what piece of information is sufficient to prove $\triangle PQO \sim \triangle NMO$?

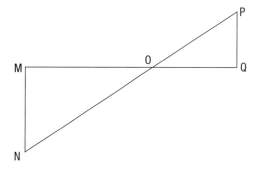

   a. O is the midpoint of side MQ.

   b. $MN \cong QP$

   c. PQ is parallel to NM.

   d. $NO \cong PO$

   e. $PQ \sim NM$

77. The school orchestra is going to randomly pick one orchestra member to go to the symphony. All orchestra students received raffle tickets based on their attendance for January. Marcia received 12 raffle tickets, Davis received $\frac{2}{3}$ the number of raffle tickets Marcia received, and Trish received two more raffle tickets than Davis. If 200 tickets were distributed in all, what is the probability that Marcia, Davis, or Trish will win?

   f. $\frac{1}{25}$

   g. $\frac{31}{200}$

   h. $\frac{3}{20}$

   j. $\frac{3}{50}$

   k. $\frac{1}{10}$

78. For all x, $(3x - 4)^2 = ?$

   a. $6x - 8$

   b. $9x^2 + 16$

   c. $9x^2 - 16$

   d. $9x^2 - 12x + 16$

   e. $9x^2 - 24x + 16$

79. There are seven bags of marbles. Bag one has two marbles. Bags two and three have six marbles each. Bag four has three marbles. Bag five has five marbles. There are four marbles in bag six, and there are nine marbles in bag seven. Which of the following statements is true?

  f. The range is less than the mean.

  g. The mean is equal to the median.

  h. The median is greater than the mode.

  j. The mode is equal to the range.

  k. The median is greater than the range.

80. Which of the following lines is perpendicular to the line $4y + 3x = 12$?

  a. $y = \frac{1}{3}x + 12$

  b. $y = -\frac{3}{4}x + 3$

  c. $y = \frac{3}{4}x - 3$

  d. $y = -\frac{4}{3}x + 4$

  e. $y = \frac{4}{3}x - 4$

81. Given m is even, $m = n + 2$, and $o = \frac{n}{2}$, which of the following statements is correct?

  f. n is odd.

  g. n is not divisible by 2.

  h. o can be an odd number.

  j. $m - n$ is odd.

  k. o is always divisible by 2

82. For all non-zero values of a and b,
$$\frac{\frac{1}{a} + \frac{1}{b}}{\frac{1}{a} - \frac{1}{b}} = ?$$

  a. $\frac{a+b}{a-b}$

  b. $\frac{a+b}{b-a}$

  c. $\frac{b-a}{a+b}$

  d. $\frac{a-b}{a+b}$

  e. $\frac{a^2 - b^2}{ab}$

83. Use the following chart. Molly owns a store in Mar Vista. For Molly's first order, she purchased 800 CDs. After she sold all of these, she placed an order for 1,200 more. After selling all of these as well, she ordered 2,100 more CDs. How much would she have saved if all the CDs had been purchased at the same time?

# CDs Purchased	Price per CD
100–999	$1.50
1,000–1,999	$1.00
2,000–2,999	$.75
3,000+	$.50

  f. $1,200

  g. $1,575

  h. $1,925

  j. $2,050

  k. $2,400

84. What is the quotient of $\frac{12x^4 + 21x^3 - 3x^2}{3x^2}$?

  a. $4x^2 + 7x$

  b. $4x^2 + 7x - 1$

  c. $9x^2 + 18x$

  d. $9x^2 + 18x - 1$

  e. $36x^6 + 62x^5 - 9x^4$

85. A navy blue jacket costs $200 and a gray jacket costs $400. If the cost of the navy blue jacket increases by 10%, and the cost of the gray jacket decreases by 5%, what will be the sum of their costs?

  f. $600

  g. $610

  h. $620

  j. $630

  k. $640

86. If $-x > y$ and $|x| > y$, which values for x and y make both statements true?

 I. $(0,0)$
 II. $(-2,3)$
 III. $(-5,4)$
 IV. $(-2,-3)$
 V. $(5,-5)$

 a. all except II
 b. all of them
 c. III and IV
 d. IV only
 e. I, II, and IV

87. Katie uses chalk to draw a circle on her driveway and the edges of the circle touch the edges of the width of her driveway. If the circumference of Katie's circle is 69 feet, and her driveway is twice as long as it is wide, approximately how long is Katie's driveway?

 f. 44 feet
 g. 9.4 feet
 h. 22 feet
 j. 32 feet
 k. 11 feet

88. A girl is 4 feet tall and casts a 6-foot shadow. How long a shadow does a tree that is 24 feet tall cast?

 a. 6 feet
 b. 36 feet
 c. 18 feet
 d. 22 feet
 e. 28 feet

89. A triangle's hypotenuse has a length of 8. If the two legs have the same length, what is this common length?

 f. $2\sqrt{2}$
 g. $6\sqrt{2}$
 h. $4\sqrt{2}$
 j. $8\sqrt{2}$
 k. $16\sqrt{2}$

90. At Smith Advertising Company, a standard 1-minute commercial can be broken down into a 5-second intro, a 40-second endorsement, and a 15-second conclusion. If Derek accidentally got a fingerprint on one frame of a 1-minute commercial, what is the probability that this frame is part of the concluding segment?

 a. $\frac{1}{12}$
 b. $\frac{3}{5}$
 c. $\frac{1}{4}$
 d. $\frac{2}{3}$
 e. $\frac{7}{15}$

91. R and S are positive integers, and the smallest positive integer that is divisible by both R and S is 120. If the greatest common factor of R and S is 3, and $R = 24$, then $S = ?$

 f. 8
 g. 14
 h. 15
 j. 18
 k. 30

92. If $\frac{6}{-y-1} = \frac{10}{-2y-3}$, what is the value of y?

 a. -4
 b. -2
 c. 2
 d. 6
 e. 16

93. A blueprint drawing has a scale of $\frac{1}{2}$-inch to 5 feet. If the blueprint shows a 2-inch by $3\frac{1}{2}$-inch rectangular room, what is the area in square feet of the actual room?

 f. 70
 g. 110
 h. 175
 j. 350
 k. 700

94. After a computer is discounted by 20%, its price is $960. What was the original price of the computer before the discount?
   a. $768
   b. $1,066
   c. $1,120
   d. $1,200
   e. $1,220

95. Five boulevards pass through both Adamstown and Bernardsville. Three avenues connect Bernardsville to Cooper Hills. Monica will travel from Adamstown through Bernardsville to Cooper Hills and back. If she does not use any of the same boulevards or avenues on the return trip, how many different routes for her entire trip are possible?
   f. 225
   g. 120
   h. 14
   j. 30
   k. 23

Answers

Paragraph 1 (U, R, S, Q, T)
Sentence **U** logically comes after the given introductory sentence. The order of sentences **R, S, Q,** and **T** is determined by their references to the claim that O'Connell Street is the widest in Europe. Sentence **R** states the Dubliners' claim and sentence **S** notes French protests against the claim. Sentence **Q** signals with the word "but" that it is refuting another statement: the French protest. Sentence **T** explains the Dubliners' argument.

Paragraph 2 (R, S, Q, U, T)
The given introductory sentence mentions the history of table forks, and sentence **R** naturally follows with a statement about what historians believe about forks. Take careful note of the chronological sequence of all the choices and their logical order becomes clear.

Paragraph 3 (T, R, U, Q, S)
Sentence **T** states what occurred "at first" after forks were introduced in England. Sentence **R** refers back to the noblemen mentioned in **T**, and gives the reason for their reaction to forks. Sentence **U** refers back to the reaction of noblemen mentioned in **T** and **R**, and states what happened next. Sentence **Q** states what happened after the fork began to catch on, and sentence **S** wraps up the paragraph.

Paragraph 4 (S, R, T, U, Q)
The given sentence introduces the topic of backdrafts and sentence **S** tells when they can occur. Sentence **R** refers back to the lack of oxygen mentioned in sentence **S**. In the rest of the choices, pay careful attention to the words and phrases that tell you what each one must follow: "Then" (sentence **T**), "That's why" (sentence **U**), and "Those other warning signs" (sentence **Q**).

Paragraph 5 (U, Q, S, T, R)
Note that sentence **S** refers back to the dean and rioting students, so it must follow sentences **U** and **Q**. The words "When she did" in sentence **Q** place it after **U**. Sentence **S** refers to Autherine Lucy's first day in class, and **T** states something that surprised her in class that day. Sentence **R** tells what she felt later.

6. b. The only information we have is where Mary is and where she is going. We do not know how Mary feels about her bus ride (choice **a**). We also do not know whether she has taken a bus ride to Boston before (choice **d**). Since we do not know how long Mary has been on the bus, we cannot assume when Mary will get to Boston (choice **c**). The passage does not include information about the return trip (choice **e**). The only information we know for certain from the two statements is that it takes approximately 4.5 hours by bus to get from New York to Boston.

7. k. Consider the facts available to us. The only thing we can conclude for certain is that Alberto found fewer albums than Jonathan, since we are told that Alberto found two, while Jonathan found three. Choices **g**, **h**, and **j** may be true, but we cannot determine their validity based on the information provided to us. Similarly, there is no way to measure the level of Jonathan and Alberto's knowledge (choice **f**).

8. e. Lake Mead is compared to Walden Pond in the first statement, and Red Swamp is compared to Brown River in the second, but no comparison is made between Red Swamp and Lake Mead. Therefore, we cannot determine whether Red Swamp is colder than Lake Mead.

9. g. The only information we can safely assume is that the store was probably already closed by the time Kevin got there, since we know it is 20 minutes away from his house and closes at midnight, and he remembered that he needed to pick up a bottle of juice at 11:50 P.M. Choice **h** is incorrect because we are told in the third statement that he did go to the store. Choices **f** and **j** could be true, but nothing in the information given supports these facts.

10. e. The city that got the least rain is in the desert. New Town is in the mountains. Last Stand got more rain than Olliopolis, so it cannot be the city with the least rain; also, Mile City cannot be the city with the least rain. Olliopolis got 44 inches of rain. Therefore, Polberg is in the desert and got 12 inches of rain. The best way to find the answer is to make a chart with the information given. You know the rainfall amounts and can match them up with their locations. Then match the locations with the towns.

RAINFALL	LOCATION	TOWN
12	Desert	P
27	Coast	M
32	Mountain	N
44	Valley	O
65	Forest	L

11. g. Refer to the chart you made for Question 10. Olliopolis got 44 inches of rain, Last Stand got 65, and Polberg got 12. New Town is in the mountains, and the city in the mountains got 32 inches of rain. Therefore, Mile City got 27.

12. a. Refer to the chart you made for Question 10. Olliopolis got 44 inches of rain, so it is not in the desert or the forest. The city in the mountains got 32 inches of rain; the coast 27. Therefore, Olliopolis is in a valley.

13. f. If no shingles are purple and all of the houses have roofs with shingles, none of the houses has a purple roof.

14. e. Although we know that Amy became ill and was taken to the hospital, we do not know whether she made other arrangements for the cats, so we can't assume that they went hungry for two days (choice **a**). We have no way of knowing whether Natasha canceled her trip (choice **b**) or how soon she returned to pick up her cats (choice **c**). Amy may have felt bad about leaving Natasha's cats (choice **d**), but we have no way of verifying this.

15. j. Although Andy is always borrowing change from May, there is no evidence that Andy took this money from May's bowl (choice **g**). We do not know whether Andy does not have an income (choice **f**) or May always has a lot of spare change (choice **h**). The only information we have based on the statements is that May keeps her spare change in a bowl on her desk.

16. d. This answer is broad enough to support all the information discussed in the passage: chemicals in the home, research on certain houseplants, the suggestion of the best plants for the job and why. Choice **a** only deals with contaminants. Choice **b** suggests our allergies are caused by chemicals in the home, when the passage suggests that we unknowingly blame our symptoms on allergies. Choices **c** and **e** suggest that the passage is only about plants in the home.

17. k. This is explicitly stated in the passage. Choice **f** is an incorrect assumption, as the passage does not discuss allergies; it states that we dismiss the symptoms, blaming allergies as the cause. Choice **g** is tempting, but it is not a specific effect of the chemicals combining; it merely states that ridding our homes of impurities seems a great task. Choices **h** and **j** are incorrect because the passage does not say that the combination of harmful chemicals triggers the process of photosynthesis or causes houseplants to die.

18. a. It is clearly stated that research has been done using *certain* houseplants. Choice **b** is incorrect because the sentence that deals with NASA suggests that *even* NASA is conducting experiments. Choice **c** reveals a faulty reading of the passage in which three of the chemicals are clearly named. Choice **d** is incorrect because the main idea of the passage is for the benefit of homeowners. Choice **e** is not mentioned in the passage.

19. g. This answer is inferred in the last line of the passage: "primal qualities . . . ability to purify their environment." Choices **f** and **k** are incorrect because *antiquity* refers to how long the species has been on the planet, which has no relationship to how long a life span the individual plants or leaves have. Choice **j** is incorrect. One cannot make a general statement about how successful the plants' reactions are in research experiments when the passage only presents us with one type of research experiment. Choice **h** is not mentioned in the passage.

20. b. The narrator informs us about the Pony Express, and then relates how he and others watched a rider go by. The narrator is clearly not on a horse (choice **a**). The narrator and the stagecoach driver (who exclaims "HERE HE COMES!") are not the same person (choice **c**). Since the narrator does see a Pony Express rider, choice **d** is wrong. A train (choice **e**) is not mentioned in the passage.

21. j. The narrator expresses excitement at seeing the rider. The passage does not support choices **f**, **g**, **h**, or **k**.

22. b. The narrator mentions "the stagecoach" and the need to "get our heads out of the windows," so he is in an enclosed, windowed, space with other people. Only choice **b** fits that description. The passage provides no support for the other choices.

23. k. Note that the question asks for the answer not included in the passage. The passage reflects all of the choices except **k**. It's just the opposite, pony-riders could travel twice as far as a stagecoach in one day.

24. d. The passage states "all that passed us and all that met us managed to streak by in the night." The other answers are not supported by the passage.

25. e. The passage describes the photograph, tells why it was important, and says that it was printed on the front page of a major newspaper, so **e** is the best choice. Choices **c** and **d** are details in the passage. Choices **a** and **b** are not covered in the passage.

26. h. Based on the passage, it is safe to assume that Parks continued to create photographic images of ordinary working people, so choice **h** is most likely. The other choices are not based on evidence in the excerpt.

27. b. Note that the question asks for the answer that is NOT supported by the passage. The passage states that the photograph was of "an African-American charwoman (a woman hired to clean) holding a broom and mop." There is no mention of other people. All the other choices *are* supported by the passage.

28. h. The text explains how Parks's photograph echoes the composition of Grant Wood's famous painting. There is no indication that the photograph is dark and mysterious (choice **f**), or that Ella Watson resembled the farmer's wife in Grant Wood's painting (choice **k**). Although Parks may have admired Grant Wood (choice **j**), the text does not support that choice, and the essay clearly shows that the images in the photograph differ from the images in the painting (choice **g**). Only choice **h** fits the information given.

29. b. The excerpt clearly states that the FSA hired photographers to communicate the plight of Americans in difficulty, so choice **b** is the best answer.

30. k. Parks calls a camera a weapon because it can create influential images like his "American Gothic." It is a weapon that contributes directly to social and political change, so choice **k** is the best answer. Although choices **f**, **g**, and **j** may be valid points, they do not fit in with Parks's quote about wanting to expose the world's evils. Choice **h** is nonsensical.

31. e. The passage is about the claims of Dennis Hope that he has the right to sell land on the moon and the arguments of those who claim he does not. Choices **a** and **b** are details from the passage, but not the main idea. Choices **c** and **d** are different positions on the argument, but neither is what the passage is about.

32. g. Dennis Hope says that he found a loophole in the 1967 United Nations Outer Space Treaty. The treaty says that nations cannot own land in outer space, but it does not say anything about individuals. Choice **f** is incorrect because the passage only says that people, not just celebrities, have a right to own moon property. Choices **h** and **k** are incorrect because these support Jakhu's argument, not Hope's argument. Choice **j** is incorrect because Hope does not say he has a right to moon property just because it will make him money. He cites the treaty as proof of his right for ownership.

33. c. The author states that "Hope is right; the rules are somewhat murky." Murky means unclear or difficult to understand. Choices **a** and **b** are details from the passage, but they don't mean that Hope might have the right to sell land. Choices **d** and **e** are not supported by the passage.

34. j. The author includes celebrity names such as Barbara Walters and George Bush. These well-known names in the community, as well as the sheer number of people who have purchased land, help to show the business's popularity. Choice **f** is incorrect because the number of people and the type of people give the idea of buying land on the moon some value. Choice **h** is incorrect because a main idea of the paragraph is the popularity of Dennis Hope's business, not whether he has the right to sell the land in space. Choice **g** and **k** are incorrect because neither is a main idea in the paragraph or even the article.

35. a. Choices **b**, **c**, and **e** are incorrect because they each refer to specific points raised in the passage, but not the entire passage. Choice **d** is too broad to represent the best title. Only choice **a** describes the point of the entire passage.

36. g. The second paragraph of the passage clearly states that Benjamin Franklin first considered the concept of DST.

37. b. The third paragraph states that the bill (which was Introduced by Sir Robert Pearce in 1909) met with great opposition, mostly from farmers.

38. j. This choice is directly supported by the fifth paragraph. The other answers refer to different laws or federal departments.

39. c. The last paragraph clearly states that during the oil embargo and energy crisis of the 1970s, President Richard Nixon extended DST through the Daylight Saving Time Energy Act of 1973 to conserve energy further.

40. b. This is the only choice broad enough to cover the entire passage. Although the other choices are also true statements, they are only details from the passage.

41. j. Note that the question asks for the issue that Anthony did NOT advocate. The question of "child care for female employees" does not appear in the passage. The other choices are issues that Anthony spoke for or did.

42. b. The passage indicates that Anthony enlisted the help of others. Although choices **c** and **d** might also be true, the passage does not emphasize those. Choice **a** is contradicted by the passage and choice **e** does not appear in the passage.

43. j. This is the best definition that describes Anthony's efforts. The paragraph indicates that she worked tirelessly for several causes. Choice **f** describes how the word crusade was used in a particular historic context. Choices **g** and **h** are too broad; choice **k** is too narrow.

44. c. The passage emphasizes Susan B. Anthony's efforts to obtain the right to vote for women, and the amendment that eventually granted that right. Although choices **a**, **d**, and **e** might be true, they are not the main idea in the passage. Choice **b** might be true, but is not mentioned in the passage.

45. g. This is the only choice that the passage supports.

46. a. **$1.30.** Calculate 8% of $16.25 by turning 8% into a decimal and then multiplying: $(0.08)(\$16.25) = \$1.30$.

47. j. **45°.** Since the vertex = (2)(base), let the measure of the base angle = b, and the measure of the vertex angle = $2b$. Then vertex + 2(base) = 180 degrees, so $2b + b + b - 180$, which simplifies to $4h = 180$, and $h = 45$ degrees.

48. c. **8.** Calculate the mean by adding the 5 pieces of data and dividing by 5: $(10 + 10 + 5 + 5 + 10) \div 5 = 8$.

49. k. **36.** Since the ratio of 5:9 is based on 14 people (5 men + 9 women = 14 people), make a proportion that models women:total. 9:14 = w:56, so $w = 36$.

50. d. Since the solution set includes the endpoints, −1.4 and 2, the circles must be filled in as well as the data points in between.

51. j. **2,800.** Let regular tickets = r and children's tickets = $(600 + r)$. The sum of children's tickets and regular tickets must equal 5,000:
$r + (600 + r) = 5,000$
$2r + 600 = 5,000$
$2r = 4,400$ and $r = 2,200$. Therefore 2,200 + 600 = 2,800 children's tickets were sold.

52. a. $x + 1$. Since $\overline{AB}$ is four times as large as $\overline{FG}$, divide $4x + 4$ to find the length of $\overline{FG}$: $\frac{4x + 4}{4} = x + 1$.

53. j. **98 m².** The diagonal of a square creates an isosceles right triangle and so the Pythagorean theorem can be written with s representing both of the legs: $s^2 + s^2 = c^2$. Substituting in the values, $2s^2 = 14^2$, $2s^2 = 196$, and $s^2 = 98$. The area of a square is also equal to s^2, so the answer is 98.

54. a. $\sqrt{58}$. The distance between two points is calculated with the formula:
$d = \sqrt{(x_2 - x_1)^2 + (y_2 - y_1)^2} = \sqrt{(1-(-2))^2 + (-3-4)^2} = \sqrt{3^2 + (-7)^2} = \sqrt{9+49} = \sqrt{58}$.

55. j. **40π cm³.** Volume of a cylinder = (area of circular base)(height) = $(\pi r^2)(h)$. $\frac{2}{3}(\pi 2^2)(15) = \frac{2}{3}(60\pi) = 40\pi$.

56. e. **58 cars.** Carla's dance squad paid $250 + $35 = $285 as expense. Since each car will pay $5 to get washed, they must wash $\frac{\$285}{\$5} = 57$ cars to cover their expenses, and 58 cars to earn more than their expenses.

57. h. $y = -2x - 2$. If you multiply the entire equation by 8, it will cancel out all of the fractions and make the equation easier to work with: $8[(\frac{1}{2}x + \frac{1}{4}y] = 8[\frac{1}{8}]$ will become $4x + 2y = 1$. Then getting y by itself, $y = -2x + \frac{1}{2}$, you can see that the slope is -2. Only answer choices **h** and **j** have a slope of -2 and only choice **h** works with the point $(0,-2)$.

58. e. 240 students. 40% of the students wanted to pursue Maternity. 40% of 1,200 = $(0.40)(1,200) = 480$. 20% of the students wanted to pursue ER. 20% of 1,200 = $(0.10)(1,200) = 240$. The difference of these two groups of students is 240.

59. j. $\frac{4}{5}$. The drawer contains a total of 20 socks, where 16 of them are *not* white. $\frac{16}{20} = \frac{4}{5}$.

60. a. $4\sqrt{\pi}$. The area of a circle = πr^2 and the area of a square = s^2. The circle has a radius of 4, so its area is 16π. If $16\pi = s^2$, then $s = 4\sqrt{\pi}$.

61. k. none of the above. None of the given statements are true: If z is negative, then $2z$ will not be > 0. Not all odd numbers are prime. $3z$ will always be even, since z is even. 2 is even and prime, not composite.

62. a. 3 sides. The sum of the interior angles of a polygon with n sides is always $180(n - 2)$. Let $180(n - 2) = 60n$.
$180n - 360 = 60n$
$120n = 360$, so $n = 3$.

63. g. angle 2 = angle 3. Angles 3 and 7 are vertical angles, so their measures are equal. Angles 4 and 8 are also vertical angles, so their measures are also equal. Because angles 4 and 7 are congruent, and angles 4 and 8 are congruent, angles 7 and 8 must be congruent. In the same way, because angles 4 and 7 are congruent and angles 3 and 7 are congruent, angles 3 and 4 must be congruent. Angles 3, 4, 7, and 8 are all congruent. In fact, because angles 3 and 4 are complementary and angles 7 and 8 are complementary (because lines CD and GH are perpendicular), all four angles measure 45°. However, nothing is known about angles 1, 2, 5, and 6. Therefore, it cannot be stated that angle 2 = angle 3.

64. e. $\frac{2}{3}x^2$ **square units.** The area of the square will be x^2. The area of the triangle will be $(\frac{1}{2})(\text{base})(\text{height}) = (\frac{1}{2})(x)(\frac{2}{3})(x) = (\frac{2}{6})x^2 = (\frac{1}{3})x^2$. Subtract the area of the triangle from the area of the square to find the area of the shaded area: $x^2 - (\frac{1}{3})x^2 = \frac{2}{3}x^2$ square units.

65. h. $49x^6y^4$. When working with exponents, a rule states that when an exponent is on the outside of parenthesis, it goes to each individual term inside the parenthesis. First, square 7 to get 49. Then, square each of the variables; this means multiplying each exponent by 2. In other words, when a power is raised to another power, simply multiply the exponents.

66. c. $\sqrt{9} + \sqrt{25} = \sqrt{64}$. $\sqrt{64} = 8$, and $\sqrt{9} + \sqrt{25} = 3 + 5 = 8$.

67. h. 120°. Since $\angle 1$ is 30° and $\angle 2$ is 90°, the measure of $\angle 3$ must be 60°. The line segment is parallel to the base of the triangle since $\angle 2$ is right, and it is a corresponding angle to the right angle at the base of the triangle. Therefore, $\angle 3$ and $\angle 4$ are also corresponding, so that means that $\angle 4$ is 60°. Since $\angle 4$ and $\angle 5$ make a straight angle, $\angle 5$ must measure 120°.

68. a. one-fourth the area of the old square. The area of a square is equal to the length of one side of the square multiplied by itself. Therefore, the length of a side of a square is s, the area of the square is s^2. If the sides of the square are halved, then the area of the square becomes $(\frac{1}{2})(s)(\frac{1}{2})(s) = (\frac{1}{4})(s^2)$. The area of the new square, $(\frac{1}{4})(s^2)$ is one-fourth the area of the old square, s^2.

69. f. –10. The definition of the function states that the term before the & symbol should be divided by the term after the & symbol, and that quotient should be added to the product of the two terms. Therefore, the value of $q\&p = \frac{q}{p} + qp$. Substitute 4 for q and –2 for p in this definition: $(\frac{4}{-2}) + 4(-2) = -2 + -8 = -10$.

70. a. $f \times g$ must be divisible by 4. $f \times g$ must be divisible by 2 and 3, and by 6, since $6 = 2 \times 3$. $f \times g$ might be divisible by 5 (for example, 30 is divisible by 2, 3, and 5), but does not have to be divisible by 4 (for example, 6 is divisible by 2 and 3, but not 4).

71. f. –278.

$V = x^3 + y[r(3s + 2) + 4r]$
$V = -2^3 + 3[-5(3(4) + 2) + 4(-5)]$
$V = -8 + 3[-5(14) + -20]$
$V = -8 + 3[-70 + -20]$
$V = -8 + 3[-90]$
$V = -8 + -270$
$V = -278$

72. c. $1,080. The lacrosse team spent 30% of their $5,400 budget on transportation: $(0.30)(\$5,400) = \$1,420$. The team's training expenses were 10% of the $5,400 budget: $(0.10)(\$5,400) = \$540$. The difference between their transportation and training expenses is $\$1,620 - \$540 = \$1,080$.

73. j. all real numbers greater than zero. The range consists of all the possible y values. Because x is squared, y will always be greater than or equal to zero.

74. d. (4,6). Plug each answer choice into the inequality; $4(6) + 6 > 3(4) + 15$ which simplifies to $24 + 6 > 12 + 15$ and finally, $30 > 27$. Since the inequality $30 > 27$ is true, choice **d** is correct.

75. k. $\frac{31}{35}$. Change the fractions so that they have common denominators, then add the numerators and keep the denominators the same: $\frac{2}{7} + \frac{3}{5} = (\frac{5}{5})(\frac{2}{7}) + (\frac{7}{7})(\frac{3}{5}) = \frac{10}{35} + \frac{21}{35} = \frac{31}{35}$.

76. c. PQ is parallel to NM. To prove $\Delta PQO \sim \Delta NMO$, you need angle-angle-angle. $\angle POQ \cong \angle NOM$ since they are vertical angles and $\angle PQO \cong \angle NMO$ since they are both right angles. If you knew that PQ was parallel to NM, then $\angle QPO$ and $\angle MNO$ would be congruent since they would be alternate interior angles.

77. h. $\frac{3}{20}$. Davis received $\frac{2}{3}$ of Marcia's 12 tickets, which is $(\frac{2}{3})(12) = 8$ tickets. Trish received $2 +$ Davis's 8 tickets = 10 tickets. Together the three of them received $12 + 8 + 10 = 30$ tickets. $\frac{30}{200} = \frac{3}{20}$.

78. d. $9x^2 - 12x + 16$. Use the acronym FOIL to remember the order of Firsts, Outsides, Insides, Lasts to expand the binomial: $(3x - 4)^2 = (3x - 4)(3x - 4) = 9x^2 - 12x - 12x + 16 = 9x^2 - 24x + 16$.

79. g. The mean is equal to the median. The median of the data set, {2, 3, 4, 5, 6, 6, 6, 9} is 5. The mode is 6. The mean is 5. The range is 7.

80. e. $y = \frac{4}{3}x - 4$. First, rewrite the equation in slope-intercept form ($y = mx + b$); $4y + 3x = 12$, $4y = -3x + 12$, $y = -\frac{3}{4}x + 3$. The slope of perpendicular lines are negative reciprocals of each other. Therefore, the slope of a line perpendicular to $y = -\frac{3}{4}x + 3$ will have a slope of $\frac{4}{3}$ since $-\frac{3}{4}$ is the slope of the given line.

81. h. ***o* can be an odd number.** If $n = 6$ and $m = 8$, then $o = \frac{6}{2} = 3$. It is possible for o to be odd.

82. b. $\frac{a+b}{b-a}$. Multiply both the numerator and the denominator by ab:

$$\frac{(ab)\left(\frac{1}{a}+\frac{1}{b}\right)}{(ab)\left(\frac{1}{a}-\frac{1}{b}\right)} =$$

$$\frac{\frac{ab}{a}+\frac{ab}{b}}{\frac{ab}{a}-\frac{ab}{b}} =$$

$$\frac{\frac{b}{1}+\frac{a}{1}}{\frac{b}{1}-\frac{a}{1}} = \frac{b+a}{b-a} = \frac{a+b}{b-a}$$

83. h. **$1,925.** Each of Molly's three orders used a different price per CD, since as the quantity goes up, the price per CD goes down. Her first order of 800 CDs cost her $800(\$1.50) = \$1,200$. Her second order of 1,200 CDs cost her $1,200(\$1.00) = \$1,200$. Her third order of 2,100 CDs cost her $2,100(\$0.75) = \$1,575$. Together these three orders cost Molly $3,975. In total, she ordered 4,100 CDs. If she bought all of these at once, she would have paid $4,100(\$0.50) = \$2,050$. Her savings would have been $3,975 − $2,050 = $1,925.

84. b. $4x^2 + 7x - 1$. Divide each term in the numerator by $3x^2$: divide the coefficient of each term by 3, and subtract 2 from the exponent of each x term since when dividing like bases, the exponents are subtracted. $\frac{12x^4}{3x^2} = 4x^2$; $\frac{21x^3}{3x^2} = 7x$; $\frac{-3x^2}{3x^2} = -1$; therefore the quotient is $4x^2 + 7x - 1$.

85. f. **$600.** Find the new cost of the navy jacket. Ten percent of $200 is $20 ($0.10 \times \$200 = \$20$). This $20 is added to the original price of $200, which makes the new price of the navy jacket $220. Next, find the cost of the gray jacket. Five percent of $400 is $20 ($0.05 \times \$400 = \$20$). Subtract this cost from the original price of $400, which makes the new price of the gray jacket $380. Last, find the sum of the costs of the two jackets. The sum will be $220 + $380 = $600.

86. c. **III and IV.** The negative sign in "$-x > y$" and the absolute value sign in "$|x| > y$" will both make −5 larger than 4 in the pair (−5,4), as well as −2 larger than −3 in the pair (−2,−3).

87. f. **44 feet.** Circumference $= 2\pi r = 69$. Since $69 = 2(3.14)r$, $r = 10.99$ feet and Katie's driveway is then 2(radius) = 22 feet wide. Since Katie's driveway is twice as long as it is wide, her driveway is 44 feet long.

88. b. **36 feet.** Set up similar triangles and a proportion to solve. $\frac{\text{Height}}{\text{Shadow}} = \frac{4}{6} = \frac{24}{\text{tree shadow}}$. Since the second numerator is 6 times larger than the first, the shadow will be 36 feet long.

89. h. $4\sqrt{2}$. Since the two legs have the same length, this is a 45-45-90 isosceles right triangle. In these triangles, the hypotenuse $= (\sqrt{2})(\text{leg})$. Therefore, $8 = (\sqrt{2})(\text{leg})$ and leg $= \frac{8}{\sqrt{2}}$ which simplifies to $\frac{8\sqrt{2}}{(\sqrt{2})(\sqrt{2})} = 4\sqrt{2}$.

90. c. $\frac{1}{4}$. Since the conclusion is 15 seconds and the commercial is 60 seconds, the probability that Derek's fingerprint is on the conclusion is $\frac{15}{60} = \frac{1}{4}$.

91. h. **15.** 3 is not a factor of 8 or 14, so **f** and **g** are incorrect. The number 120 is not evenly divisible by 18, so **j** is incorrect. The greatest common factor of 24 and 30 is 6 rather than 3, so **k** is incorrect. Only 15, choice **h**, is divisible by 3 and has the greatest common factor of 3 with 24.

92. a. −4. Cross multiply and solve for y: (6)
$(−2y − 3) = (10)(−y − 1); −12y − 18 =$
$−10y − 10; −2y − 18 = −10; −2y = 8,$ so
$y = −4.$

93. k. 700. Set up proportions to find the
dimensions of the actual room:
$\frac{0.5 \text{ in}}{5 \text{ ft}} = \frac{2 \text{ in}}{w \text{ ft}}$ and $\frac{0.5 \text{ in}}{5 \text{ ft}} = \frac{3.5 \text{ in}}{l \text{ ft}}$
Cross-multiply and solve: $0.5w = 10$ and
$0.5l = 17.5; w = 20$ ft and $l = 35$ ft. Multiply
the actual length by the actual width to give
an actual area: $20 \times 35 = 700$ square feet.

94. d. $1,200. Use the relationship "Discounted
Price = Original Price − Amount of
Discount." Let p = the Original Price. Then
the discount would equal $(0.20)(p)$ and this
can be used in the equation as follows:
$\$960 = p − (0.20)(p)$. Therefore, $\$960 =$
$1p − (0.20)(p) = 0.80p$ and $p = \$1,200.$

95. g. 120. Monica can take 5 different roads from
Adamstown to Bernardsville, then three
different roads to Cooper Hills. On her
return, since she will not use the same
roads, she can take 2 roads from Cooper
Hills to Bernardsville, and 4 roads from
there back to Adamstown. Multiply these
numbers together to see how many different
routes are possible: $5 \times 3 \times 2 \times 4 = 120.$

ADDITIONAL ONLINE PRACTICE

W hether you need help building basic skills or preparing for an exam, visit the LearningExpress Practice Center! Using the code below, you'll be able to access additional SHSAT online practice. This online practice will also provide you with:

Immediate Scoring
Detailed answer explanations
Personalized recommendations for further practice and study

Log in to the LearningExpress Practice Center by using the URL: **www.learnatest.com/practice**

This is your Access Code: **7762**

Follow the steps online to redeem your access code. After you've used your access code to register with the site, you will be prompted to create a username and password. For easy reference, record them here:

Username: _____   **Password:** _____

With your username and password, you can log in and access your additional practice materials. If you have any questions or problems, please contact LearningExpress customer service at 1-800-295-9556 ext. 2, or e-mail us at **customerservice@learningexpressllc.com**

NOTES

NOTES